JOURNAL OF CYBER SECURITY AND MOBILITY

Volume 6, No. 3 (July 2017)

JOURNAL OF CYBER SECURITY AND MOBILITY

Aim

Journal of Cyber Security and Mobility provides an in-depth and holistic view of security and solutions from practical to theoretical aspects. It covers topics that are equally valuable for practitioners as well as those new in the field.

Scope

The journal covers security issues in cyber space and solutions thereof. As cyber space has moved towards the wireless/ mobile world, issues in wireless/mobile communications will also be published. The publication will take a holistic view. Some example topics are: security in mobile networks, security and mobility optimization, cyber security, cloud security, Internet of Things (IoT) and machine-to-machine technologies.

Published, sold and distributed by:
River Publishers
Alsbjergvej 10
9260 Gistrup
Denmark

River Publishers
Lange Geer 44
2611 PW Delft
The Netherlands

Tel.: +45369953197
www.riverpublishers.com

Journal of Cyber Security and Mobility is published four times a year.
Publication programme, 2017: Volume 6 (4 issues)

ISSN 2245-1439 (Print Version)
ISSN 2245-4578 (Online Version)
ISBN 978-87-93609-72-3 (this issue)

JOURNAL OF CYBER SECURITY AND MOBILITY COMMUNICATIONS

Volume 6, No. 3 (July 2017)

ZHENYAN LIU, YIFEI ZENG, YIDA YAN, PENGFEI ZHANG
AND YONG WANG / Machine Learning for Analyzing Malware   227–244

FÉLIX IGLESIAS VÁZQUEZ, ROBERT ANNESSI AND TANJA
ZSEBY / Analytic Study of Features for the Detection of Covert
Timing Channels in Network Traffic   245–270

AHMAD MK NASSER, DI MA AND PRIYA MURALIDHARAN /
An Approach for Building Security Resilience in AUTOSAR
Based Safety Critical Systems   271–304

DANG NGUYEN, DAT TRAN, WANLI MA AND
DHARMENDRA SHARMA / Random Number Generators
Based on EEG Non-linear and Chaotic Characteristics   305–338

MASOOMEH SEPEHRI, ALBERTO TROMBETTA AND
MARYAM SEPEHRI / Secure Data Sharing in Cloud Using
an Efficient Inner-Product Proxy Re-Encryption Scheme   339–378

Machine Learning for Analyzing Malware

Zhenyan Liu, Yifei Zeng, Yida Yan,
Pengfei Zhang and Yong Wang

Beijing Key Laboratory of Software Security Engineering Technology,
School of Software, Beijing Institute of Technology,
Beijing 100081, China
E-mail: zhenyanliu@bit.edu.cn; 453624629@qq.com;
1415654264@qq.com; zhpngfei@163.com; wangyong@bit.edu.cn

Received 17 November 2017; Accepted 22 November 2017;
Publication 15 December 2017

Abstract

The Internet has become an indispensable part of people's work and life, but it also provides favorable communication conditions for malwares. Therefore, malwares are endless and spread faster and become one of the main threats of current network security. Based on the malware analysis process, from the original feature extraction and feature selection to malware analysis, this paper introduces the machine learning algorithms such as classification, clustering and association analysis, and how to use these machine learning algorithms to effectively analyze the malware and its variants.

Keywords: Malware analysis, Machine learning, Classification, Clustering, Association analysis.

1 Introduction

Malicious software, or *malware*, plays a part in most computer intrusions and security incidents. Any software that does something that causes harm to a user, computer, or network can be considered malware, including viruses, trojan horses, worms, rootkits, scareware, and spyware [1].

Journal of Cyber Security, Vol. 6_3, 227–244.
doi: 10.13052/jcsm2245-1439.631

In recent years, due to the widespread of malwares in the network, the number of network security incidents is increasing year by year, the relevant statistics show that the number of network security incidents caused by malwares increased by more than 50% per year from the 1990s [2]. These network security incidents not only reflect the vulnerability of system and network security, but also lead huge losses to the current development based on Internet infrastructure. They may cause economic losses to the network users and businesses, even lead to network paralysis.

There are two fundamental approaches to malware analysis: static and dynamic. *Static analysis* involves examining the malware without running it, while *Dynamic analysis* involves running the malware in virtual or real environment [1]. Both approaches analyze the malware based on different features. *Static analysis* uses internal code features and *Dynamic analysis* uses external behavior features on the system or in the network. *Static analysis* is straightforward and can be quick, but it can miss important behaviors. *Dynamic analysis* can capture some important behaviors, but it won't be effective with all malware. Therefore, it is a more common solution how to use *static analysis* together with *dynamic analysis* in order to completely analyze suspected malware.

The number of current malwares is very large and new malwares appear faster and faster, thus traditional detection technology have been unable to cope with the current malware detection because of their speed and efficiency. On the other hand, the traditional method has high maintenance costs, and need a large number of manual experience to carry out sample analysis and extract the rules [3]. In recent years, machine learning for analyzing malware has been widely recognized, which can effectively make up the traditional methods [4–7].

In particular, we first present the framework for analyzing malware by machine learning in Section 2, which is an infrastructure for our review of analyzing malware by machine learning. And then, in Section 3, we provide a critical review of the innovative research developments targeting the malware analysis problem by machine learning, including classification, clustering, and association rules. Finally, we make a prospect on the focus of future research of malware analysis by machine learning in Section 4.

2 The Framework for Analyzing Malware by ML

Machine learning for analyzing malware is still a newcomer, but it has already obtained the huge success in fact. Commonly used machine learning algorithms are clustering, classification and association analysis and so on. The classification technology can not only be used to detect unknown malware before starting its malicious behavior, but also to label malware family according the exact or fuzzy features of malware [8]. Clustering and association analysis are more useful in how to make family decisions and homology analysis of unknown samples, thus to increase the speed of handling or improve the efficiency of manual analysis.

The framework for analyzing malware by machine learning is shown in the Figure 1.

In this framework there are two subsequent phases that the first phase is to obtain the features of the malware including internal code features and external behavior features on the system or in the network, and then the second phase is to build a machine learning model for analyzing the malware based on these features of the malware.

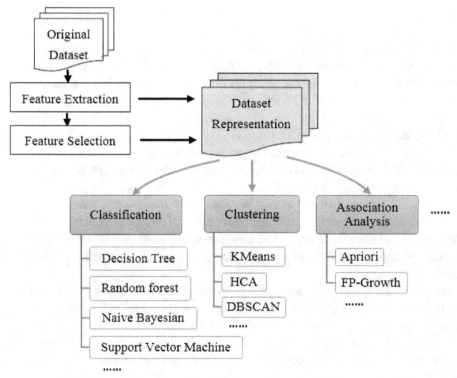

Figure 1   The framework for analyzing malware by machine learning.

The quality of the feature has a greater impact on the performance of the machine learning model, so the first phase in this framework is very critical [9]. In this phase, feature acquisition is generally divided into two steps, one is *original feature extraction*, and the other is *feature selection*. But according to different practical task requirements, sometimes the *original feature extraction* is only used, and sometimes the *feature selection* is also incorporated into this phase.

The main task of the *original feature extraction* is to extract various types of features. Many studies focused on internal code features of the malware such as byte n-grams [5, 6], OpCode [10–12] and PE (Portable Executable) features [13, 14]. In addition, a few researchers utilized external behavior features on the system such as creating file, hiding service, opening port [15], and some works also devoted to employ external behavior features in the network such as DNS (Domain Name System) answer and TTL (Time To Live) value [16].

However, many features from *original feature extraction* are generally redundant, many of which not contributing to build machine learning model (or even degrading its performance). So it is very necessary to select the most important and indispensable features (i.e. *feature selection*). *Feature selection* aims to select a subset of features which are the best for building machine learning model [17]. There are various *feature selection* methods employed in malware detection field, including Document Frequency [18], Information Gain [19], and chi-square [20].

In Kaggle [21], feature engineering was a very key part, which directly determined the game's performance. In Kaggle 2015, we can see some first-class participants used the frequency of the PE header file, the designated dll, opcode, call to select features, it's also worth mentioning that the third place used the L1 regularization of SVM (Support Vector Machine) to remove roughly some irrelevant features, and then utilized the *feature_importances* parameter of Random Forest to make a better choice.

Based on the features obtained, the original samples are represented, and then the representative vectors are the input for a learning model. Next step is the second phase in this framework which is to build a machine learning model, such as classification, clustering and association analysis. In the following section, we will introduce in detail the leading algorithms related to machine learning for analyzing malware.

3 Machine Learning Algorithms

Various machine learning algorithms, including classification, clustering and association analysis have been commonly used for analyzing malware. The most popular classification algorithms are Decision Tree, Random forest, Naive Bayesian, Support Vector Machine, etc. In clustering, we often see KMeans, HCA (hierarchical cluster analysis) and DBSCAN (Density-Based Spatial Clustering of Applications with Noise). In association analysis, Apriori and FP-Growth (Frequent Pattern-Growth) are frequently used.

3.1 Classification

Classification is a supervised technique which is usually divided into two phases: The first phase is to train a classifier using classification algorithm based on the labeled samples, and then the subsequent phase applies this classifier to assign a class to each new data item [22]. The training data and the test data are represented on the basis of the same feature space. A number of classification algorithms applied to analyze malware can be found in the literature. In this section, we will introduce these classification algorithms in detail.

3.1.1 Decision tree

Decision tree algorithms construct a model of decisions made based on actual values of attributes in the data. Each non-leaf node in the tree (including the root node) corresponds to a test of a non-category attribute in the training sample set, and a test result for each branch of the non-leaf node, and each leaf node represents a class or class distribution. A path from the root node to the leaf node forms a classification rule [23]. Decision tree algorithm has many variants, such as ID3 (Iterative Dichotomiser 3) and C4.5, but the basis is similar. C4.5 algorithm has been used more in the field of malware detection.

Zhu et al. [24] proposed a new method for detecting unknown malwares under the Windows platform: the API (Application Program Interface) function dynamically invoked with PE file as the research object, using the sliding window mechanism to extract the feature, and adopt the decision tree C4.5 Algorithm to detect unknown malwares. The results showed that the decision tree C4.5 was higher than the minimum distance classifier and the Naive Bayesian algorithm, and the stability of C4.5 was better than the other two algorithms.

Zhao et al. [16] developed a new system to detect APT (Advanced Persistent Threat) malware which relied on DNS to locate command and control servers. In this system, a classifier of *Malicious DNS Detector* was trained by a series of feature related DNS, which employed J48 decision tree algorithm and obtained the true positive rate of 96.3% and the false positive rate of 1.7%.

Perdisci et al. [25] presented a novel system called FluxBuster for detecting and tracking malicious flux network via large-scale passive DNS analysis. The system, using hierarchical clustering algorithm to group the domains and C4.5 classifier to classify these clusters, was capable of accurately detecting previously unknown flux networks days or weeks in advanced before they appear in blacklists.

3.1.2 Random forest

The Random Forest is an ensemble learning method that constructs a multitude of decision trees and outputs a prediction that is the mode of the classes of the individual trees. A subset of the training dataset (local set) is chosen to grow individual trees, with the remaining samples used to estimate the goodness of fit. Trees are grown by splitting the local set at each node according to the value of a random variable sampled independently from a subset of variables [26].

Tian et al. [27] presented a malware classification technique based on string information using several well-known classification algorithms, including Random Forest, IB1(Instance Based 1), AdaBoost. The experiments showed that the IB1 and Random Forest classification methods were the most effective for this domain.

Zhao et al. [28] defined the features according to both the statistical information and the topology of the function call graph, and used the J48, Bagging, Random Forest and other algorithms to detect viruses, Trojans and worms. The comparative experiments showed that Random Forest was better than other algorithms with 96% accuracy.

Shabtai et al. [29] used the DF method to select different quantities of OpCode N-grams as a feature set, and then used Random Forest, ANN(Artificial Neural Network), SVM, Naive Bayesian and other methods, finally reported that Random Forest was the best algorithm with more than 95% correct rate.

3.1.3 Naive Bayesian

The Naive Bayesian classifier is based on Bayes' theorem, which provides a way of calculating the posterior probability, and works on assumption called class conditional independence. A Naive Bayesian model is easy to build, but often does surprisingly well and is widely used because it often outperforms more sophisticated classification methods.

Zhu et al. [30] proposed a parameter valid window model for system call which improved the ability of operation sequence to describe behavior similarity. On this basis, they presented a malware classification approach based on Naive Bayesian and parameter valid window. The experiment results showed that this approach was effective, and the performance and accuracy of training and classification were improved through parameter valid window.

Sayfullina et al. [31] presented a scalable and highly accurate method for malware classification based on features extracted from APK (Android application package) files. They explored several techniques for tackling independence assumptions in Naive Bayes and proposed Normalized Bernoulli Naive Bayes classifier that resulted in an improved class separation and higher accuracy.

Passerini et al. [32] developed a system named FluXOR which could detect and monitor fast-flux service networks. FluXOR was divided into three components: collector, monitor and detector. The detector fed the set of collected features of the suspicious hostname to the naive Bayesian classifier for the classification.

3.1.4 Support vector machine

Built on structural risk minimization and VC dimension theory of Statistical Learning, the basic classification idea of SVM is to determine the optimal separating hyperplane, and to find this hyperplane can be converted to solve an equivalent quadratic optimization problem. The result of the optimization process is a set of coefficients to determine the optimal separating hyperplane.

Li et al. [33] studied a malware detection scheme for Android platform using an SVM-based approach, which integrated both risky permission combinations and vulnerable API calls and used them as features in the SVM algorithm. The experiments showed that the proposed malware detection scheme was able to identify malicious Android applications effectively and efficiently.

Mcgrath et al. [34] presented a method of examining fast flux, DNS flux and double flux in the phishing context through classification algorithm that Support Vector Machines, using the features such as the number of IP address and associated ASN that collected various information from each URLs.

By analyzing the patterns of the DNS queries from the fast-flux botnets, Yu et al. [35] extracted six features for constructing the weighted SVM in order to distinguish the normal network domain access from the fast-flux botnet domain access. The evaluation suggested that the approach was effective in detecting the fast-flux botnets.

3.2 Clustering

Clustering is an unsupervised technique which can group data using some measure of inherent similarity between instances, in such a way that samples in one cluster are very similar (compactness property) and samples in different clusters are different (separateness property). After the clusters are created, a labeling phase takes care of annotating each cluster with the most descriptive label [36]. Numerous clustering algorithms applied to analyze malware are available in the literature. In this section, we will introduce these clustering algorithms in detail.

3.2.1 KMeans

KMeans is very popular for cluster analysis. KMeans aims to partition n samples into k clusters in which each sample belongs to the cluster with the nearest mean, serving as a prototype of the cluster. KMeans is very simple and can be easily implemented in solving many practical problems. There exist a lot of extended versions of K-Means such as XMeans [37].

Sharma [38] grouped the executables on the basis of malware sizes by using Optimal KMeans Clustering algorithm and used these obtained groups to select promising features for training (Random Forest, J48, etc) classifiers to detect variants of malwares or unknown malwares.

Dietrich et al. [39] proposed a mechanism to detect DNS C&C (Command and Control) in network traffic and described a real-world botnet using DNS C&C. The mechanism used k-Means clustering and Euclidean Distance based classifier with the feature related DNS, which could obtain higher true positive rates.

Antonakakis et al. [40] proposed a novel detection system called Pleiades to identify DGA(Domain Generation Algorithm) bots through monitoring traffic below the local recursive DNS server and analyzing NXDomain

(Non-Existent Domain) responses. In Pleiades, Xmeans clustering algorithm was used to group domain subsets into larger clusters.

3.2.2 HCA

HCA is a method of cluster analysis which seeks to build a hierarchy of clusters. Strategies for HCA generally fall into two types: Agglomerative and Divisive. The agglomerative HCA is a "bottom up" approach which each sample starts in its own cluster, and pairs of clusters are merged as one moves up the hierarchy. And the divisive HCA is a "top down" approach which all samples start in one cluster, and splits are performed recursively as one moves down the hierarchy [41].

Perdisci et al. [42] presented a network-level behavioral malware clustering system using *single-linkage agglomerative hierarchical clustering*, which focused on HTTP-based malware and clustered malware samples based on a notion of structural similarity between the malicious HTTP traffic they generated.

Chatzis et al. [43] proposed a method that using *divisive hierarchical clustering* to divide machines into worm-infected machines and non-infected machines according to the similarities in their DNS communication patterns. And the FP rates and FN rates remained under 1% and 6% in overall email worm detection.

Based on DGA domains and NXDomain traffic, Thomas et al. [44] constructed a system that computed pairwise DNS similarity measures and subsequently performed *single-linkage agglomerative hierarchical clustering*. The resulting clusters grouped the DGA domains to distinctive clusters based on their DNS traffic and contained only domains relevant to that particular variant.

3.2.3 DBSCAN

DBSCAN is a density-based clustering algorithm: given a set of points in some space, it groups together points that are closely packed together (points with many nearby neighbors), marking as outliers points that lie alone in low-density regions (whose nearest neighbors are too far away) [45].

To distinguish the variations of malicious code, Qian et al. [46] studied the malicious behavior of malwares, then computed the similarity of characteristics and the call graphs which were extracted by disassembly tools. They employed DBSCAN to discover the family of malicious code.

Schiavoni et al. [47] presented a mechanism called Phoenix based on the DBSCAN clustering algorithm, which could not only tell DGA- and non-DGA-generated domains apart using a combination of string and IP-based features, characterize the DGAs behind them, but also find group of DGA-generated domains that are representatives of the respective botnets.

Xiao et al. [48] proposed a behavior-based malware cluster method inspired by the characteristic that Android malwares in a same family behaved similarly. With collected feature data, DBSCAN was employed to cluster malwares into various families. The evaluation results showed that the accuracy rate of malware clustering can reach 91.3% at best, and the correct prediction rate is 82.3% for evaluated samples.

3.3 Association Analysis

Finding implicit relationships between items from large-scale data set is called association analysis or association rule learning. Transactions may have a certain degree of law and relevance, malware is the same, between the behavior and family, there have a certain relationship, so you can tap the useful association rules for the malware analysis. There are many algorithms for mining association rules cited in the literature. In this section, we will introduce these algorithms in detail.

3.3.1 Apriori

Apriori is a classical algorithm which can mine frequent itemsets for Boolean association rules. Apriori employs an iterative approach known as a *level-wise* search, where k-itemsets are used to explore (k+1)-itemsets. To improve the efficiency of the *level-wise* generation of frequent itemsets, an important property called the Apriori property, i.e. all nonempty subsets of a frequent itemset must also be frequent, is used to reduce the search space [49].

Studying the rules of API calling sequence of some viruses and their variants in the implementation process, Zhang et al. [50] used Apriori algorithm to extract valuable association rules from the known virus API call sequence. And according to balancing the false positive rate and false negative rate, the best confidence threshold to guide the virus detection was acquired.

In [51], malware code was disassembled and preprocessed into sequential data, an Apriori-like algorithm was used to discover sequential pattern and remove normal pattern, and the result pattern set can be used to detect unknown malware. Experimental result showed that the method had high accuracy rate and low false positive rate.

Adebayo et al. [52] presented a classification system of Android malware using candidate detectors generated from Apriori algorithm improved with particle swarm optimization to train three different supervised classifiers. In this method, features were extracted from Android applications byte-code through static code analysis, selected and used to train supervised classifiers.

3.3.2 FP-Growth

The Apriori algorithm needs to scan the database once every iteration, generate a large number of candidate sets, spend a lot of time on the I/O, and the algorithm is inefficient. FP-growth algorithm [53] only needs to scan the database twice, which greatly speeds up the algorithm. Therefore, the application of FP-Growth algorithm in malware analysis is more extensive and in-depth.

In order to identify malicious web campaigns, Kruczkowski et al. [54] constructed a FP-SVM system which employed the FP-growth algorithm and the SVM method. This system could be successfully used to analyze a huge amount of dynamic, heterogenous, unstructured and imbalanced network data. In this system, the FP-growth algorithm was applied to produce the training dataset.

In [55], the frequent itemsets were obtained by the FP-Growth algorithm which extracted the URLs of the malware events, then mined the association rules from them, linked all the associated rules to the graphs, finally used the modular approach to divide the graph into different groups and analyzed a malware web site map.

In [56], a weighted FP-Growth frequent pattern mining algorithm was proposed for malicious code forensics. Different API call sequences were assigned different weights according to their threaten degree to obtain frequent patterns of serious malicious codes and more accurate analysis results. Compared with the original FP-Growth algorithm, the weighted algorithm can obtain higher accuracy when used for evidence analysis.

From the above analysis, we can see that all kinds of methods in different scenarios have their own advantages, and there are no absolute merits of the points. The performance of the algorithm often depends on the application of the scene, as well as the selected features, or the characteristics of the samples. And no algorithm can have any advantage in any application scenario, so the combination of the actual scene is the most critical. Moreover, it is worth mentioning that the original feature extraction and feature selection are very critical, which determine the final performance of the machine learning model.

4 Conclusion

Malware analysis has become a hot topic in recent years in network security. The emergence of plenty of new malwares makes the traditional analysis method no longer completely effective. How to extract the most representative features of malware, and further to maximize the speed and accuracy of malware analysis still need to intensive study. Therefore, it is necessary to use a more efficient and intelligent approach – machine learning, to detect and analyze unknown malwares.

The future development of malware analysis by machine learning will focus on three major research areas. Firstly, the existing machine learning algorithm should be improved and combination with specific application. Secondly, the improvement of the feature extraction and feature selection method should be considered to be expanded based on the semantic level, so that we can get more accurate features. The third more valuable research direction can be considered is to make a combination of feature and instance selection methods to study the scalability of malware database.

Acknowledgements

This work was financially supported by National Key R&D Program of China (2016YFB0801304).

References

[1] Sikorski, M., and Honig, A. (2012). *Practical Malware Analysis: The Hands-On Guide to Dissecting Malicious Software*. San Francisco, CA: no starch press.

[2] Liao, G., and Liu, J. (2016). A malicious code detection method based on data mining and machine learning. *J. Inf. Secu. Res.* 2, 74–79.

[3] Huang, H. X, Zhang, L., and Deng, L. (2016). Review of Malware Detection Based on Data Mining. *Computer Sci.* 43, 13–18.

[4] Lee, D. H., Song, I. S., and Kim, K. J., et al. (2011). A Study on Malicious Codes Pattern Analysis Using Visualization. In *International Conference on Information Science and Applications. IEEE Computer Society*, 1–5.

[5] Kolter, J. Z., and Maloof, M. A. (2006). Learning to Detect and Classify Malicious Executables in the Wild. *J. Mach. Learn. Res.* 6, 2721–2744.

[6] Schultz, M. G., Eskin, E., and Zadok, E., et al. (2000). Data Mining Methods for Detection of New Malicious Executables. Security and

Privacy, 2001. S&P 2001. In *Proceedings. 2001 IEEE Symposium*. IEEE, 38–49.

[7] Lai, Y. (2008). A Feature Selection for Malicious Detection. In *Ninth Acis International Conference on Software Engineering, Artificial Intelligence, Networking, and Parallel/distributed Computing. IEEE Computer Society,* 365–370.

[8] Mao, M., and Liu Y. (2010). Research on Malicious Program Detection Based on Machine Learning. Software Guide, 9, 23–25.

[9] Domingos, P. (2012). A Few Useful Things to Know about Machine Learning. *ACM*, 55, 78–87.

[10] Karim, M. E., Walenstein, A., and Lakhotia, A., et al. (2005). Malware Phylogeny Generation Using Permutations of Code. *J. Computer Virology*, 113–23.

[11] Bilar, D. (2007). Opcodes as Predictor for Malware. *Int. J. Electronic Security & Digital Forensics*, 1, 156–168.

[12] Santos, I., Brezo, F., and Ugarte-Pedrero, X., et al. (2013). Opcode Sequences as Representation of Executables for Data-Mining-Based Unknown Malware Detection. Information Sciences, 231, 64–82.

[13] Perdisci, R., Lanzi, A., and Lee, W. (2008). Classification of Packed Executables for Accurate Computer Virus Detection. *Pattern Recognition Letters* 29, 1941–1946.

[14] Ding, Y., Yuan, X., and Tang, K., et al. (2013). A Fast Malware Detection Algorithm Based on Objective-Oriented Association Mining. *Computers & Security*, 39, 315–324.

[15] Lu, Y. B., Din, S. C., and Zheng, C. F., et al. (2010). Using multi-feature and classifier ensembles to improve malware detection. *J. Chung Cheng Institute of Technology*, 39, 57–72.

[16] Zhao, G., Xu, K., and Xu, L., et al. (2015). Detecting APT malware infections based on malicious DNS and traffic analysis. *IEEE Access,* 3, 1132–1142.

[17] Liang, C. (2012). Research on the Main Techonologies in Malware Code Detection. Yangzhou University.

[18] Moskovitch, R., Feher, C., and Tzachar, N., et al. (2015). Unknown Malcode Detection Using OPCODE Representation. In *Intelligence and Security Informatics, First European Conference, EuroISI 2008*, Esbjerg, Denmark, 204–215.

[19] Kolter, J., and Maloof, M. (2006). Learning to detect and classify malicious executables in the wild. *J. Mach. Learn. Res.* 7, 2721–44.

[20] Siddiqui, M., Wang, M. C., and Lee, J. (2008). Data Mining Methods for Malware Detection Using Instruction Sequences. In *IASTED International Conference on Artificial Intelligence* and Applications. ACTA Press, 358–363.

[21] http://www.kaggle.com/malware-classification

[22] Fang, Z. (2011). Research and Implementation of Malware Classification. National University of Defense Technology.

[23] Li, W. (2010). Research and Implementation of Mobile Customer Churn Prediction Based on Decision Tree Algorithm. Beijing University.

[24] Zhu, L. J., and XU, Y. F. (2013). Application of c4.5 algorithm in unknown malicious code identification. *J. Shenyang University Chemical Technol.* 27, 78–82.

[25] Perdisci, R., Corona, I., and Giacinto, G. (2012). Early detection of malicious flux networks via large-scale passive dns traffic analysis. *IEEE Transactions on Dependable & Secure Computing* 9, 714–726.

[26] Mistry, P., Neagu, D., Trundle, P. R., et al. (2016). Using random forest and decision tree models for a new vehicle prediction approach in computational toxicology. *Soft. Comput.* 20, 2967–2979.

[27] Tian, R., Batten, L., and Islam, R., et al. (2010). "An automated classification system based on the strings of trojan and virus families", in *International Conference on Malicious and Unwanted Software*. IEEE, 23–30.

[28] Zhao, Z., Wang, J., and Wang, C. (2013). An unknown malware detection scheme based on the features of graph. *Secu. Commun. Netw.* 6, 239–246.

[29] Shabtai, A., Moskovitch, R., and Feher, C., et al. (2012). Detecting unknown malicious code by applying classification techniques on opcode patterns. *Secu. Inf.* 11.

[30] Zhu, K., Yin, B., and Mao, Y., et al. (2014). Malware classification approach based on valid window and naive bayes. *J. Computer Res. Dev.* 51, 373–381.

[31] Sayfullina, L., Eirola, E., Komashinsky, D., Palumbo, P., Miche, Y., Lendasse, A., and Karhunen, J. (2015). Efficient detection of zero-day Android malware using normalized bernoulli naive bayes. In *Trustcom/BigDataSE/ISPA, IEEE*, 1, 198–205. IEEE.

[32] Passerini, E., Paleari, R., and Martignoni, L., et al. (2008). FluXOR: Detecting and monitoring fast-flux service networks. In *Proceedings of the 5th international conference on Detection of Intrusions and Malware, and Vulnerability Assessment*, 186–206.

[33] Li, W., Ge, J., and Dai, G. (2015). Detecting malware for android platform: An svm-based approach. In *Cyber Security and Cloud Computing (CSCloud), 2015 IEEE 2nd International Conference,* 464–469. IEEE.

[34] McGrath, D. K., Kalafut, A., and Gupta, M. (2009). Phishing infrastructure fluxes all the way. *IEEE Security & Privacy.*

[35] Yu, X., Zhang, B., Kang, L., and Chen, J. (2012). Fast-flux botnet detection based on weighted svm. *Inf. Technol. J.* 11, 1048–1055.

[36] Ceri, S., Bozzon, A., Brambilla, M., Della, V. E., Fraternali, P., and Quarteroni, S. (2013). Web Information Retrieval. Springer Berlin Heidelberg.

[37] Pelleg, D., and Moore, A. W. (2000). X-means: Extending K-means with Efficient Estimation of the Number of Clusters. In *ICML,* 1, 727–734.

[38] Sharma, A. (2015). Grouping the Executables to Detect Malwares with High Accuracy. In *International Conference on Information Security and Privacy,* 667–674.

[39] Dietrich, C. J., Rossow, C., and Freiling, F. C., et al. (2012). On Botnets that Use DNS for Command and Control. In *Seventh European Conference on Computer Network Defense.* IEEE, 9–16.

[40] Antonakakis, M., Perdisci, R., and Nadji, Y., et al. (2012). From Throwaway Traffic to Bots: Detecting the Rise of DGA-based Malware. In Usenix Conference on Security *Sympo*sium. 24–40.

[41] *Hierarchical Clustering.* Available at: https://en.wikipedia.org/wiki/Hierarchical_clustering

[42] Perdisci, R., Ariu, D., and Giacinto, G. (2013). Scalable fine-grained behavioral clustering of http-based malware. *Computer Netw.* 57, 487–500.

[43] Chatzis, N., Popescu-Zeletin, R., and Brownlee, N. (2009). Email worm detection by wavelet analysis of DNS query streams. In *Computational Intelligence in Cyber Security, 2009. CICS'09. IEEE Symposium on* 53–60. IEEE.

[44] Thomas, M., and Mohaisen, A. (2014). Kindred domains: detecting and clustering botnet domains using DNS traffic. In *Proceedings of the 23rd International Conference on World Wide Web,* 707–712. ACM.

[45] Density-based spatial clustering of applications with noise (DBSCAN). Available at: https://en.wikipedia.org/wiki/DBSCAN

[46] Qian, Y., Peng, G., and Wang, Y., et al. (2015). Homology analysis of malicious code and family clustering. *Computer Engineering & Applications* 51, 76–81.

[47] Schiavoni, S., Maggi, F., Cavallaro, L., and Zanero, S. (2014). Phoenix: DGA-based botnet tracking and intelligence. In *International Conference on Detection of Intrusions and Malware, and Vulnerability Assessment* 192–211. Springer, Cham.

[48] Xiao, Y., Su, H., Qian, Y., and Peng, G. A., (2016). Behavior-based family clustering method for android malwares. *Journal of Wuhan University: Natural Science Edition*, 62, 429–436.

[49] Han, J., and Kamber, M., (2011). *Data Mining Concept and Techniques.* Elsevier.

[50] Zhang, W., Zheng, Q., and Shuai, J. M., et al. (2008). New malicious executables detection based on association rules. *Computer Eng.* 34, 172–174.

[51] Wang, X. Z., Sun, L. C., and Zhang, M., et al. (2011). Malicious Behavior Detection Method Based on Sequential Pattern Discovery. *Computer Eng.* 37, 1–3.

[52] Adebayo, O. S., and AbdulAziz, N. (2014). Android malware classification using static code analysis and apriori algorithm improved with particle swarm optimization. In *Information and Communication Technologies (WICT), 2014 Fourth World Congress* 123–128. IEEE.

[53] Han, J., Pei, J., and Yin, Y. (2000). Mining frequent patterns without candidate generation. In *ACM Sigmod Record* 29, 1–12. ACM.

[54] Kruczkowski, M., Niewiadomska-Szynkiewicz, E., and Kozakiewicz, A. (2015). FP-tree and SVM for Malicious Web Campaign Detection. In *Asian Conference on Intelligent Information and Database Systems* 193–201. Springer, Cham.

[55] Li-xiong, Z., Xiao-lin, X., Jia, L., Lu, Z., Xuan-chen, P., Zhi-yuan, M., and Li-hong, Z. (2015). Malicious URL prediction based on community detection. In *Cyber Security of Smart Cities, Industrial Control System and Communications (SSIC), 2015 International Conference* 1–7. IEEE.

[56] Li, X., Dong, X., and Wang, Y. (2013). Malicious code forensics based on data mining. In *Fuzzy Systems and Knowledge Discovery (FSKD), 2013 10th International Conference* 978–983. IEEE.

Biographies

Zhenyan Liu works at School of Software, Beijing Institute of Technology. She received her Ph.D. degree in Computer Architecture from Institute of Computing Technology Chinese Academy of Sciences, China. Her current research interests include big data, artificial intelligence, and cyber security. She's a fellow of the China Computer Federation (CCF).

Yifei Zeng is a M.Sc. student at Beijing Institute of Technology since autumn 2017. He attended the Guangdong Ocean University of Zhanjiang, China where he received his B.Sc. degree in Software Engineering in 2016. His M.Sc. work focuses on cyber security and artificial intelligence.

Yida Yan is a M.Sc. student at Beijing Institute of Technology since autumn 2016. She attended the Hebei Normal University, China where she received her B.Sc. degree in Software Engineering in 2015. Her M.Sc. work focuses on data mining and cyber security. She has obtained Data Mining Senior Engineer Certificate, in China.

Pengfei Zhang is a M.Sc. student at Beijing Institute of Technology since autumn 2017. He attended the Nanjing University of Science & Technology, China where he received his B.Sc. degree in Computer Science in 2015. He has been a member of TP-Link from 2015 to 2016. His M.Sc. work focuses on cyber security and artificial intelligence.

Yong Wang works at School of Software, Beijing Institute of Technology. She received her Ph.D. degree in Computer Science from Beijing Institute of Technology, China. Her current research interests include cyber security and machine learning. She's a fellow of the China Computer Federation (CCF).

Analytic Study of Features for the Detection of Covert Timing Channels in Network Traffic

Félix Iglesias Vázquez, Robert Annessi and Tanja Zseby

CN Group, Institute of Telecommunications, TU Wien, Austria
E-mail: felix.iglesias@nt.tuwien.ac.at; robert.annessi@tuwien.ac.at;
tanja.zseby@nt.tuwien.ac.at

Received 30 November 2017; Accepted 3 December 2017;
Publication 18 December 2017

Abstract

Covert timing channels are security threats that have concerned the expert community from the beginnings of secure computer networks. In this paper we explore the nature of covert timing channels by studying the behavior of a selection of features used for their detection. Insights are obtained from experimental studies based on ten covert timing channels techniques published in the literature, which include popular and novel approaches. The study digs into the shapes of flows containing covert timing channels from a statistical perspective as well as using supervised and unsupervised machine learning algorithms. Our experiments reveal which features are recommended for building detection methods and draw meaningful representations to understand the problem space. Covert timing channels show high histogram-distance based outlierness, but insufficient to clearly discriminate them from normal traffic. On the other hand, traffic features do show dependencies that allow separating subspaces and facilitate the identification of covert timing channels. The conducted study shows the detection difficulties due to the high shape variability of normal traffic and suggests the implementation of semi-supervised techniques to develop accurate and reliable detectors.

Keywords: Covert timing channels, Network traffic analysis, Classification, Anomaly detection, Feature selection.

Journal of Cyber Security, Vol. 6_3, 245–270.
doi: 10.13052/jcsm2245-1439.632

1 Introduction

Covert channels are parasitic communication channels, which are exploited to hide the existence of actual communications but for the senders and the receivers of the covert information. They are parasitic because they are built on top of other systems, usually related to technology or information transmission means. In the case of TCP/IP protocols, covert channels use either TCP/IP header fields or time properties of network traffic packets to hide the secret message.

The existence of covert channels is usually—but not necessarily—related to illegal and criminal activities, e.g., data leakage, malware propagation, penetration attacks. One example is the network attacks during December 2015 on Ukraine power companies, whose computers were infected with Gcat. The Gcat malware created covert channels on Gmail applications to conceal C&C (command-and-control) operations and to pass unperceived by security systems [8]. Covert channels are a serious problem related to security and system vulnerabilities; not in vain, they were already identified as a security threat for communication networks almost 40 years ago [25].

This work focuses on covert channels over TCP/IP and, specifically, on covert timing channels, i.e., whenever time properties of TCP/IP transmissions are exploited. We continue the investigations started with [12], where covert channels were classified by studying the implications and challenges faced from the detection side. Additionally, in [12] a general solution for covert channel identification called DAT (from Descriptive Analytics of Traffic) is proposed. The DAT methodology is based on representing traffic flows with a set of statistical measurements and estimations. In [13] and [15] the scope was reduced to timing channels, analyzing and testing the detection with supervised classification in [13] and with unsupervised algorithms in [15]. Even in spite of the fact that covert timing channels are *anomalies* from a semantic understanding, from the perspective of statistical properties they remain in zones also occupied by normal traffic, yet they can be differentiated by supervised analysis (as shown in [15]). In other words, the separation boundaries between traffic flows with and without covert timing channels are not located in low density areas of the problem space (at least when analyzing the spaces created by the features studied in [15]).

In this paper, we extend the experiments and improve the cited research in [12, 13, 15] as follows:

1. *Study of new features for randomness and predictability*
   The suitability of five new coefficients related to the estimation of randomness and predictability in time series is studied. The addition of new

features to the basic DAT vector format increases computational costs but is expected to enhance detection performance, being therefore justified in environments that demand high security and deal with assumable traffic volumes.

2. *Validation with novel techniques*

A validation phase was added to the experiments. In the validation phase, in addition to the eight covert timing techniques used in previous experiments, detectors are evaluated versus two new covert timing channels published in 2017, which are not present during training and testing phases. The robustness of statistical detection frameworks is therefore here tested as well as the hypothesis that suggests that different covert timing techniques are prone to similarly exploit the timing capacities of network traffic (within the scope of the features under test).

The rest of the paper is organized as follows: Section 2 briefly describes the covert timing channel techniques used in the experiments. Section 3 describes the features under study. Section 4 depicts the experimental design and Section 5 discusses the obtained results. Finally, conclusions are drawn in Section 6.

2 Explored Covert Timing Channel Techniques

For the experiments we have selected ten covert timing channel techniques. Eight of them are popular in the field of covert channel detection, widely described in [13] and in the original papers (we provide the corresponding references). The last two techniques have been published recently in [7]. We use them here for validation and testing detectors against unknown/novel techniques. We abbreviate technique names with three capital letters (or two and one number) that refer to the authors who published them. Techniques are briefly described later in this section, and Table 1 lists the specific parameters used for the generation of covert channels.

For a better understanding of the developed methods, it is important to clarify two different terms related to time properties. The Inter Arrival Time (IAT) refers to the time between packets seen from the perspective of the receiver; on the other side, the Inter Departure Time (IDT) is the time between packets seen from the perspective of the sender. Synthetically expressed: IAT = IDT + tx_delay, where "tx delay" is the transmission delay.

2.1 Packet Presence (CAB)

In this technique communication partners are time synchronized and agree on a predefined time interval t_{int}. The presence or absence of a packet within the predefined time interval stands for the covert symbol 1 or 0 respectively [4].

2.2 Differential/Derivative (ZAN)

This is a technique originally described for the Time-to-Live (TTL) field [33], yet easily applicable to IATs. This technique makes use of two parameters, t_b and t_{inc}. t_b refers to the base IDT at which packets may be sent and t_{inc} to the time that may be added or substracted from t_b. If the covert symbol 0 is to be transmitted, the IDT is set to t_b; if the covert symbol 1 is to be transmitted, however, the IDT is set to $t_b \pm t_{inc}$. Addition and substraction are applied alternately on subsequently occurring 1 symbols.

2.3 Fixed Intervals (BER)

This straightforward technique [1] agrees on two different IDTs to mask binary symbols. For instance, t_0 for 0 and t_1 for 1.

2.4 Jitterbug/Modulus (SHA)

The technique [28] is designed to interfere legitimate communications. It uses a base sample interval ω and adds some delay to IDTs. A covert 1 or a 0 is interpreted depending on if a given IAT is divisible by ω or only by $\omega/2$.

2.5 Huffman Coding (JIN)

By using Huffman coding, every covert symbol is encoded in a set of packets with different IDTs [32]. The proposed codification tries to optimize communication bandwidth based on the frequency of ASCII symbols observed in English texts. The given implementation only covers a set of basic lower case characters and numbers.

2.6 One Threshold (GAS)

This technique [9] uses a time threshold t_h to discriminate 0s and 1s. If a given IAT is above t_h it will mark 1, 0 if below. For the implementation we used additional times to create the covert channel and match the proposed rule (Section 1).

2.7 Packet Bursts (LUO)

In this technique [20] packets are sent in bursts. The number of packets sent per burst marks the intended covert symbol to transmit. Between two subsequent bursts, some time t_w is waited. In principle, this technique is not exploited for binary channels, but manages some symbols (few, about 10 or 16).

2.8 TCP Timestamp Manipulation (GIF)

This technique [10] uses TCP timestamps to convey covert information. Packet IDTs are altered in such a way that the least significant bit (LSB) of the TCP timestamp matches the covert information. TCP timestamps are updated at a specific clock frequency, usually 100, 250, or 1000 Hz. For this reason, this technique establishes a waiting t_b between consecutive packets (authors suggest a minimum of 10 ms). If the LSB is the same as the desired numerical covert symbol, the packet is sent; if not, the sender checks the LSB again after a defined t_w.

2.9 ASCII Binary Encoding (ED1)

This technique [7] encodes the covert symbol 0 with a waiting time t_0 (authors suggest 300 ms), whereas the covert symbol 1 triggers the immediate sending of a packet. Note that t_0 is not defining any IDT; therefore, sending 0s does not imply sending any packet. For example, if 'A', which corresponds to the ASCII encoding '01000001', is to be covertly send, the sender will follow this sequence: (1) wait 300 ms, (2) send one packet, (3) wait 1.5 s (300 ms×5), (4) send one packet. Messages are forced to finish transmitting a '1'.

2.10 5-Delay Encoding (ED2)

This technique [7] encodes every letter of the English alphabet with a 5-digit numerical code where each digit corresponds to a different IDT. Authors propose a waiting time between covert letters, t_w.

3 Selected Features

As mentioned in Section 1, a recent classification of covert channels is presented in [12]. Moreover, DAT is introduced as a general methodology to identify covert channels in network traffic flows. The DAT approach is fundamentally based on a set of statistical figures and estimations extracted from TCP/IP flow header fields as well as IATs. The specific field under study—IAT for the case of covert timing channels—is initially tackled from a time series analysis perspective. Further derived studies that focus on covert timing channels are developed in [13] and [15]. Here we explore the features used in these three cited works ([12, 13] and [15]) and add five new features related to regularity and randomness in time and symbol series.

Before depicting features, it is worth remembering that detectors must define two time parameters to enable time series analysis:

- *Sampling time*, which states the desired binning (time resolution, or granularity) for the time series analysis. DAT detectors and the experiments conducted here work with a *1-millisecond* sampling time by default.
- *Observation period*, which establishes the maximum time-window for the observation of a given flow. DAT detectors and the experiments conducted here work with a *5-minute* observation period by default. The feature vector extracted from IATs of analyzed flows is:

$$\text{flow\_vec} = \{U, S_k, \mu_{\omega S}, S_s, p(Mo), \rho_A, H_a, T_R, T_S, H_q, K, pkts\} \tag{1}$$

The meaning of each feature is described below.

3.1 Simple Statistical Figures

- U — *number of unique values.* U stores the number of non-repeated IAT values observed in the analyzed flow.
- *pkts* — *total number of packets in the flow.* *pkts* is simply a counter that contains the total number of packets observed in the analyzed flow.
- *p(Mo)* — *Mode frequency.* *p(Mo)* is a percentage value calculated as the quotient between the number of IAT Mode value repetitions and the total observed IATs in the flow (i.e., *pkts* – 1).
- *c* — *estimation of potential covert byte-equivalent symbol.* The calculation of *c* is based on empirical tables and tries to estimate the number of potential covert bytes sent in a flow [12]. It uses S_k (defined in Section 3.2) to guess the number of different symbols that the covert channel might be applying. For example, if S_k reveals a possible binary channel, *c* becomes: *c* = (*pkts* –1)/7, assuming that 7 IATs values are necessary to transmit a covert byte-equivalent symbol[1].

---

[1]It might be surprising the use of "7" instead of "8" in the quotient for estimating byte-equivalent symbols. The *c* index is an inflated estimation of the transmitted covert information and assumes that it consists of text. In this respect, note that ASCII most relevant/used symbols are less than $2^7 = 128$.

An example should help to understand the introduced features. For instance, given the following sequence of IATs in a flow *i*:

$$\text{IAT}_i = \{5, 15, 14, 4, 5, 6, 16, 15, 5, 15, 14, 15, 4, 5\}\text{ms} \tag{2}$$

Feature values would be: $U = 6$, i.e., the cardinality of the unique values set: $\{4, 5, 6, 14, 15, 16\}$; $p(Mo) = 4/14 = 0.29$, given that the Mode (15) occurs 4 times; *pkts* = 15; and $c = 14/7 = 2$, since $S_k = 2$.

3.2 Multimodality Estimation

Multimodality refers to values that either significantly occur more often than others in the series or appear as attractors in the series distribution. Estimating multimodality means providing the number of such attractors. The features related to multimodality are:

- S_k— *multimodality based on kernel density estimations.* S_k is the number of peaks of the curve that approximates the empirical distribution of IATs by using a gaussian kernel. Figure 1 shows two examples with similar estimated curves and both with $S_k = 2$. Kernel density estimation for assessing multimodality is widely described in [29], and their application for covert channels is discussed in [12].
- S_s — *multimodality based on pareto analysis.* S_s is the number of outstanding values that appear in the IAT histogram. It follows the principles of Pareto analysis [24] and the specific calculation is depicted in [12]. In short, it is the number of values that show a *considerable* high frequency compared to the rest of the histogram. Figure 1 shows two examples: in the left one $S_s = 2$, whereas in the right plot $S_s = 6$. S_s is expected to be irrelevant for the case of IATs as—considering a millisecond resolution—IAT series distributions tend to look like distributions of continuous variables.
- $\mu_\omega S$ — *average distribution width.* $\mu_\omega S$ is the mean of the characteristic standard deviations of the gaussians drawn by the kernel density estimations. More intuitively: it captures the average peak width of the mountains that appear in the empirical distribution.

3.3 Regularity/predictability Estimators

The coefficients described in this section try to provide an estimation of the randomness, the unpredictability or the regularity of a time series. ρ_A was

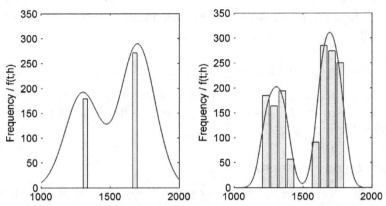

Figure 1 Histograms (bars) and kernel density estimations (curves) of two different numerical series. Left plot: $S_k = 2$, $S_s = 2$. Right plot: $S_k = 2$, $S_s = 6$. Density magnitudes are omitted as only the number of peaks is relevant for the purpose of the estimation. Figure taken from [12].

initially proposed in [12], whereas the other five measurements are added here to test if their inclusion enhances detector performances. In addition to the associated computational costs, the main drawback of all these coefficients is that they require time series be *long enough* (the minimum threshold is approximately placed between 10 and 100 elements). For short time series, coefficient values are either meaningless or not computable. This fact does not imply a serious problem for covert timing channel detection as, by default, short flows—specially in binary covert channels—cannot contain significant amounts of information and can be directly discarded by the detector. For instance, a binary channel that hides the sentence "hello world!" as 7-bit ASCII characters requires $1 + 12 \times 7 = 85$ packets (84 IATs).

- ρ_A— *sum of autocorrelation coefficients.* ρ_A is the sum of autocorrelation coefficients, i.e., the IAT time series is paired with delayed versions of itself and correlation coefficients are extracted. Later, such coefficients are summed and a final value is obtained. In short, ρ_A tries to unmask repetition patterns that fit the studied delays. Values close to 1 disclose the existence of patterns and regularity.
- T_R — *runs test.* The runs test [2] checks if it is reasonable to consider that each element in the studied time series is independent and originates from the same distribution. For a non-binary time series A, an intermediate step is required:

$$B = A - \mu_A \tag{3}$$

where μ_A is the statistical mean of A. R is the *runs of B*, i.e., the number of sub-series of negative or positive values within B. Finally,

$$Z = \frac{R - E[R]}{S[R]} \tag{4}$$

where $E[R]$ is the expected number of *runs,* and $S[R]$ stands for the expected standard deviation of R. Finally,

$$T_R = |Z| - Z_{1-\alpha/2} \tag{5}$$

For time series with more than 20 elements, they are considered non-random if T_R is positive.[2] This test is widely explained in [2].

- T_S — *sign test.* The sign test [21] is similar to the run test. Given the time series A, instead of counting *runs,* it takes consecutive pairs of values and constructs a new time series as follows:

$$B = A_2 - A_1, A_3 - A_2, \ldots, A_n - A_{n-1} \tag{6}$$

being n the total number of elements in A. P is the *signs* of B, i.e., the number of positive elements in B. Therefore,

$$Z = \frac{P - E[P]}{S[P]} \tag{7}$$

where $E[P]$ is the expected number of *signs,* and $S[R]$ stands for the expected standard deviation of R. T_S is defined similarly to T_R:

$$T_S = [Z] - Z_{1-\alpha/2} \tag{8}$$

where time series with more than 20 elements are considered non-random if T_S is positive.

- K — *Kolmogorov complexity or compressibility.* Given a string, the Kolmogorov complexity is defined as the length of the shortest computer program that generates such string [18]. The calculation of the Kolmogorov complexity, as originally defined, presents problems related to computability. Nevertherless, considering long-enough strings, using lossless compression for approximating Kolmogorov complexity has proven to satisfactorily estimate upper bounds in different domains [19]. In our experiments, we approximate K as:

$$K = \text{len}(B)/\text{len}(A) \tag{9}$$

---

[2]For a 5% significance level, $Z_{1-\alpha/2} = 1.96$.

with len(A) being the length of the time series A and len(B) the length of the compressed version of A by $zlib^3$, i.e., B = zlib(A).

- **H_q** — *Hurst exponent.* The Hurst exponent is strongly linked to the *fractal* properties of a sequence of values or series, therefore it also provides a measure of autocorrelation and *self-similarity*. In the case of time series, it gives a quantitative estimation of the trend to regress to the mean value or to be inclined to move in a certain direction [17]. Values between $0.0 < H_q < 0.5$ and $0.5 < H_q < 1.0$ stand respectively for negative and positive autocorrelations, while $H_q = 0.5$ corresponds to uncorrelated Brownian processes [6].[4]

- **H_a** — *Approximate Entropy.* Approximate entropy is a statistical method devised to measure complexity and regularity of a system, which has proven to be suitable for short (but more than 100 data points) and noisy time-series [27].[5] H_a reflects the existence of patterns in a series that makes future values more predictable. Small H_a values tending to 0 are expected for time series that contain repetitive patterns, whereas high values are expected for chaotic behaviours.

As an example, Figure 2 shows two time series: T1, generated at random, and T2, exhibiting a clear pattern. Regularity/predictability coefficients are shown below the figure.

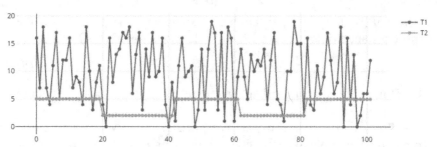

Figure 2   Regularity/predictability coefficients for two example time series.
T1: $\rho_{A_{1.5}} = 0.73, T_R = -1.76, T_S = -1.61, K = 1.00, H_q = 0.68, H_a = 2.32.$
T2: $\rho_{A_{1.5}} = 0.09, T_R = 7.49, T_S = 14.66, K = 0.31, H_q = 0.71, H_a = 0.08.$

---

[3]https://zlib.net/

[4]For the experiments we have used the *hurst* implementation from Python library *nolds 0.3.4*: https://pypi.python.org/pypi/nolds

[5]For the experiments we have used the *approximate entropy* implementation from Python library *nolds 0.3.4*: https://pypi.python.org/pypi/nolds

4 Design of Experiments

The experiments conducted in this research use real data flows for the traffic free of covert channels (henceforth called *overt datasets*) and flows with covert channels created by an ad-hoc traffic generation framework (henceforth called *overt datasets*). We describe them in this section.

As introduced in Section 3, flows are tracked 5 minutes as longest (*observation period*), and the *sampling time* for the IAT data is fixed to 1 millisecond. Final preprocessed datasets for experiment replication and further algorithm testing are publicly available to download from our webpage in [31].

4.1 Overt Datasets

Real traffic—in principle assumed free of covert channels—has been obtained from the MAWI Working Group Traffic Archive[6]. The MAWI project promotes network traffic research by daily publishing 15 minutes of TCP/IP backbone traces, from 2:00 pm to 2:15 pm GMT. MAWI datasets are anonymized and do not contain payload. We used captures from three days in 2017: a) January 31, b) February 28, and c) March 31, randomly selecting flows to create three overt datasets, namely: *nocc_training, nocc_testing and nocc_validation*.

Since, for the case of covert timing channels, DAT detectors automatically consider flows with less than 10 packets as overt flows (i.e., they are too short to contain a covert timing channel), all flows with less than 10 packets were removed from the overt dataset. In [13] and [15] flows with less than two packets were removed instead. This difference makes current experiments more demanding and challenging as covert flows are only compared with overt flows that potentially look like covert flows. Performance indices are expected to be worse due to this reason. It is important to remark here that datasets used in the experiments do not try to be representative of real traffic conditions (were the rate of covert channels is absolute negligible), but submit the analysis to problem spaces that are suitable for the knowledge extraction and feature testing.

After preprocessing steps, overt datasets contain 30000 flows in *nocc_training*, 30000 flows in *nocc_testing* and 46193 in *nocc_validation*.

---

[6]http://mawi.wide.ad.jp/mawi/

4.2 Covert Datasets

Covert datasets were also generated in three groups, namely:

- *cc_training* (1022 flows), including flows generated by using CAB, BER, SHA, JIN, GAS, LUO, ZAN and GIF techniques.
- *cc_testing* (1036 flows), including flows generated by using CAB, BER, SHA, JIN, GAS, LUO, ZAN and GIF techniques.
- *cc_validation* (1232 flows), including flows generated by using CAB, BER, SHA, JIN, GAS, LUO, ZAN, GIF, ED1 and ED2 techniques.

Each covert dataset was generated with a different set of files to be secretly sent. Each set of files consisted of various types of data, including plain-text (text files, list of passwords, technical reports and programming scripts), images (PNG and JPEG), compressed files (in ZIP and GZ formats), and encrypted files (in 3DES and GnuPG formats). Each singular channel was generated with a different parameterization, random seed and data to be covertly sent. Parameter value ranges were defined according to the original publications (whenever provided); otherwise, they were tuned based on traffic measurement expert knowledge. We also increased some parameter values whenever the source publication did not properly consider transmission delays for large networks. Table 1 shows the value ranges used for the random generation of parameters.

Table 1 Parameters used for the generation of covert channels. Parameters randomly fell within the shown intervals (uniform random distribution)

Technique	Parameters
tx_delay	By default, transmissions delays are modeled with a Lomax (Pareto Type II) distribution with $\alpha = 3$ ms and $\lambda = 10$ ms.
CAB	$t_{int} \in [60,140]$ ms, being t_{int} the base time window in which the presence or absence of a packet sets the covert symbol.
BER	$t_0 \in [10, 50]$ ms, $t_1 \in [80, 220]$ ms. t_0 stands for the IDT of covert 0s and t_1 for covert 1s.
SHA	The ground transmission was modeled by using a Gamma distribution with $k \in [40, 760]$ ms and $\phi \in [40, 360]$ ms. The SHA technique uses $\omega \in [10, 90]$ ms as sampling interval to manipulate the sending with little delays.
GAS	IDT distribution for 0-packets: $t_0 = th - t_s$ IDT distribution for 1-packets:: $t_1 = th - t_s + t_a$, $th \in [100, 300]$ ms, $t_s \in [60,140]$ms, $t_a \in [20, 80]$ ms. $th > t_a + t_a$
JIN	We used the codification proposed in [32].
LUO	$t_w \in [50, 250]$ ms. t_w is the waiting time between bursts. Packets in a burst are sent every ms.

ZAN	$t_b \in [30, 70]$ ms, $t_{inc} \in [20, 40]$ ms. t_b is the base IDT between packets. t_{inc} is added, subtracted or nor applied based on the covert symbol and the previous IDT.
GIF	The minimum time between packets is $t_b \in [10, 30]$ ms. The time to recheck TCP timestamps is $t_w \in [4,12]$ ms.
ED1	Waiting time for zeros is $t_0 \in [200, 400]$ ms.
ED2	Waiting time between covert symbols is $t_w \in [130,170]$ ms. The 5-code is built by translating ASCII decimal values into 5-base equivalent numbers. 5-code IDTs: $t_0 \in [11,19]$ ms, $t_1 = t_0 + t_{inc}$ ms, $t_2 = t_1 + t_{inc}$ ms, $t_3 = t_2 + t_{inc}$ ms, $t_4 = t_3 + t_{inc}$ ms, with $t_{inc} \in [11,19]$ ms. The specific numbers have been chosen to deal with path delay variation such that the probability for decoding errors is reduced.

4.3 Analysis Methods

Covert and overt datasets were explored in sequential steps:

1. *Univariate analysis and feature correlation*
   As a first step, features were studied separately for the *cc_training* and *nocc_training* datasets by univariate analysis, aiming to detect noticeable differences in simple statistical figures. Later on, Pearson correlation between features were checked in order to detect distinct feature dependencies in overt and covert traffic.

2. *Feature selection*
   cc_training and nocc_training were joined in a single *training* dataset to study feature dependencies with regard to the binary class labels. Features were weighted by feature selection filters and hybrid schemes with different criteria. i.e., decision trees, maximum relevance, information gain, correlation and the gini index—the theoretical background of such indices can be consulted in [26] and [23]. Additionally, feature weighting methods were embedded in a subsampling structure to reinforce result robustness by means of *stability selection* [22]. Obtained ranks were compared and a final feature set is proposed.

3. *Binary classification and validation*
   In this phase, *training* datasets are presented to learners. Obtained models are tested with *testing* datasets and validated with *validation* datasets, which explore the effect of including non-trained covert timing channels in the analysis. The training is performed with 10-fold cross-validation, with stratified sampling to keep label proportions in each validation fold. Experiments are repeated with different feature sets to evaluate the results of the feature selection phase. Used learning schemes are: decision trees,

random forests, SVMs, neural networks, Bayes-based ensembles and k-Nearest neighbors classifiers.

4. *Unsupervised analysis*

   Training data is also analyzed by unsupervised outlier detection algorithms in order to see if the new proposed features make covert flows be outliers. Experiments slightly differ from the ones conducted in [15], where overt and covert datasets where presented together to the algorithms. Such experimental scheme might break the normal environment where covert channels appear by overpopulating the space with covert samples (in [15] they consisted on about 5% of the total flows). Covert channels are expected to be much more infrequent among overt data. Therefore, here, outlier detection models are calculated without covert channels, and covert flows are later contrasted with the model and ranked one by one, i.e., every covert flow is isolated and independently compared with the whole overt dataset. Used algorithms are: LOF [3], COF [30], INFLO [16] and HBOS [11]. Used performance indices are: *P@n*, precision at the top n ranks; *Adj.P@n*, adjusted P@n; *AP*, average precision; *Adj.AP*, adjusted AP; *MaxF1*, Maximum F1 score; *Adj.MaxF1*, adjusted MaxF1 score; *ROC/AUC*, area under the ROC curve. Explanations of such indices can be consulted in [5].

5 Results

The results of the experiments described in Section 4.3 are shown and discussed in this section.

5.1 Univariate Analysis and Feature Correlation

Correlation analysis shows some linear dependencies among the studied features in overt traffic (Figure 3, left plot). Even thought it obviously depends on the features under observation, the characteristics of network traffic make finding high-related features a common situation, as observed in [14]. The positive correlation between the number of U (i.e., unique IATs values) and S_k (i.e., distribution-based multimodes) is not surprising, as well as the correlation between *pkts* and c, T_R or T_S, which are expected to be higher as the number of packets increase. A special attention deserves indices among the regularity estimators. Entropy measures H_a show inverse correlation with U, which makes sense according to the definition of entropy. The Kolmogorov coefficient K shows inverse correlation with $p(Mo)$ and T_S, which also makes sense because K inversely defines chaos or irregularity if compared to $p(Mo)$ or T_S.

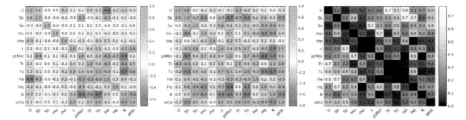

Figure 3   Results of the correlation analysis. Left matrix (A): correlations in the overt dataset. Middle matrix (B): correlations in the covert dataset. Right matrix (C): absolute differences between the correlations in the overt and the covert data set, i.e., $c_{i,j} = |a_{i,j} - b_{i,j}| \forall i, j$ being i and j indices for rows and columns, and $a_{i,j}, b_{i,j}, c_{i,j}$ elements of A, B, C respectively.

The plot in the middle of Figure 3 shows the case of the covert dataset. Covert traffic shows more extreme correlation indices. This fact suggests that covert flows follow more strict and regular structures with respect to the selected features. Another way to look at this picture is acknowledging that overt traffic is richer in shapes and possibilities. Such results foresee a classification scenario where covert flows can be discriminated, but false positives might be also likely. The right plot in Figure 3 shows which features—high values, light backgrounds—might be potentially determinant to establish classification boundaries to separate overt and covert traffic.

The comparison between univariate analysis of features for overt and covert datasets reveal significant differences in the central tendency measures (Table 2). However, values that significantly differ from overt to covert traffic also show high dynamic ranges (represented by the standard deviation).

Table 2   Univariate statistics for overt and covert traffic. Since feature distributions are not Gaussians, we use nonparametric Confidence Intervals (approximately 95%) over the median

	Overt Traffic					Covert Traffic				
	mean	stdev	median	CI_low	CI_high	mean	stdev	median	CI_low	CI_high
u	28.75	38.73	15.00	15.00	15.00	196.10	96.17	159.00	155.00	165.00
S_k	18.83	28.88	11.00	11.00	11.00	168.38	114.49	139.00	132.00	146.00
S_s	2.33	6.61	2.00	2.00	2.00	2.62	2.16	2.00	2.00	2.00
$\mu_{\omega S}$	0.01	0.05	0.00	0.00	0.00	0.02	0.04	0.01	0.01	0.01
ρ_A	0.11	0.08	0.10	0.10	0.10	0.04	0.03	0.04	0.03	0.04
c	88.53	1462.21	3.29	3.29	3.29	781.63	2139.75	289.36	246.71	351.00
$p(Mo)$	0.27	0.22	0.20	0.20	0.20	0.07	0.07	0.05	0.05	0.05
T_R	−0.17	3.57	−1.05	−1.07	−1.02	1.36	4.57	−0.39	−0.50	−0.22
T_S	3.21	18.86	0.00	0.00	0.00	1.87	4.71	0.29	0.10	0.50
H_a	6.95	4.43	10.00	10.00	10.00	0.57	0.06	0.58	0.58	0.58
H_q	129.77	335.41	0.41	0.41	0.41	1.20	0.44	1.29	1.26	1.31
K	0.97	0.12	1.00	1.00	1.00	1.00	0.02	1.00	1.00	1.00
pkts	414.49	4014.59	23.00	22.00	23.00	3647.22	4465.61	1774.00	1615.00	2062.00

This fact anticipates overlapping areas in the problem space that can hinder accurate classification.

In short, univariate analysis and correlation tests reveal:

- Overt and covert traffic show different central tendencies and feature-correlation profiles. This is an evidence of the existence of patterns that can be learned by classification schemes.
- Some feature pairs—specially $[U, H_\alpha], [U, T_R], [S_k, H_q], [c, S_s],$ $[p(Mo), \mu_{\omega S}]$—show opposed correlation relationships when overt and covert traffic are compared. Such pairs might be keys that facilitate the binary classification.
- Redundancy is high and common in features, specially for covert traffic. This fact can imply difficulties for selecting the right features and for some classification techniques (e.g., naive Bayes).

5.2 Feature Selection

Feature selection experiments expose that features are ranked differently depending on the used feature selection method, as shown in Figure 4. This is not surprising due to the redundancy among features observed during correlation analysis. Nevertheless, all feature selection methods seem to agree in neglecting ρ_A and H_q whereas emphasizing K and c.

The preponderance of K can surprise due to the fact that its correlation relationships among features do not significantly differ when overt and covert traffic are compared each other. In any case, such peculiarity does not imply that K cannot be correlated with the class label. But K is indeed not *linearly* correlated with the class label ($\rho = 0.04$), and univariate statistics in Table 2

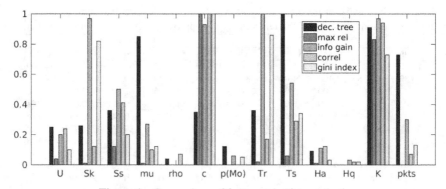

Figure 4   Comparison of feature selection methods.

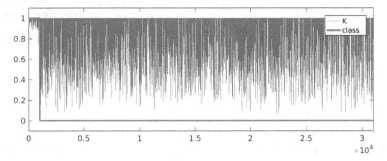

Figure 5   *K* values throughout covert and overt sample flows.

do not show differences between covert and overt traffic for *K*. Nevertheless, *K* exhibits a noticeable behaviour change when values are visually compared from an overall perspective (Figure 5).

Summarizing, based on the feature selection analysis we can cluster features in four groups, which remark their importance for distinguishing between covert and overt traffic:

- High relevant: K and c.[7]
- Medium relevant: S_k, S_s, T_R, T_S
- Low relevant: U, $\mu_{\omega S}$, H_a, $p(Mo)$, *pkts*.
- Negligible: ρ_A and H_q.

As observed in [13], where features are selected by decision trees, c is stated as a decisive feature and ρ_A is neglected. However, with regard to the previous work the importance of other features vary due to the inclusion of the new regularity and randomness features, derived redundancies and the influence of new feature combinations. Also the selection of overt datasets with longer flows, whose characteristics are more similar to covert flows, has an effect on classifiers and feature selection algorithms, which are forced to give solutions with finer granularity.

5.3 Binary Classification and Validation

Based on the feature selection analysis, classification experiments are run to verify the suitability of the selected features. Also, covert channels based on two new techniques are incorporated in a validation step to see if models drawn in training are able to detect novel covert timing techniques, meaning

---

[7]It's worth remembering that c is a function of S_k and *pkts*.

that new covert channels are likely to manipulate traffic capacities in a similar way (from a statistical perspective). Classification is performed with decision tree cores over three different feature sets:

- Feature Set A: high, medium, low and negligible features (all).
- Feature Set B: high, medium and low relevant features.
- Feature Set C: high and medium relevant features.

Tables 3, 4 and 5 show performances for every feature set when decision trees are used to create classification models. The revision of such tables disclose some findings:

- The performance downgrade when using feature set B (high, medium and low relevant) if compared with the feature set A (all) is minor, likely related to specific non-generalizable cases. From a general perspective H_q and ρ_A can be considered inefficient for detecting covert timing channels and, therefore, removed.

Table 3   Experiment performance with Feature Set A (all)

	Training		Testing		Validation	
	PC	PO	PC	PO	PC	PO
RC	1001 (TP)	21 (FN)	986 (TP)	50 (FN)	1036 (TP)	196 (FN)
RO	40 (FP)	29960 (TN)	41 (FP)	29959 (TN)	96 (FP)	46097 (TN)
Acc.(C):	99.80% ± 0.12%		99.71%		99.38%	
Prec.(C):	96.20% ± 2.24%		96.01%		91.52%	
Recall(C):	97.95% ± 1.75%		95.17%		84.09%	
AUC(C):	0.962 ± 0.099		0.973		0.903	

PC: predicted covert, PO: predicted overt, RC: real covert, RO: real overt,
TP: true positive, TN: true negative, FP: false positive, FN: false negative,
(C): covert as positive class, Acc.: accuracy, Prec.: precision, AUC: area under ROC curve.

Table 4   Experiment performance with Feature Set B (excluded negligible)

	Training		Testing		Validation	
	PC	PO	PC	PO	PC	PO
RC	997 (TP)	25 (FN)	946 (TP)	74 (FN)	991 (TP)	241 (FN)
RO	37 (FP)	29963 (TN)	96 (FP)	29961 (TN)	81 (FP)	46112 (TN)
Acc.(C):	99.80% ± 0.08%		99.54%		99.32%	
Prec.(C):	96.45% ± 1.65%		96.10%		92.44%	
Recall(C):	97.56% ± 1.60%		92.86%		80.44%	
AUC(C):	0.959 ± 0.100		0.973		0.924	

PC: predicted covert, PO: predicted overt, RC: real covert, RO: real overt,
TP: true positive, TN: true negative, FP: false positive, FN: false negative,
(C): covert as positive class, Acc.: accuracy, Prec.: precision, AUC: area under ROC curve.

Table 5   Experiment performance with Feature Set C (excluded negligible and low relevant)

	Training		Testing		Validation	
	PC	PO	PC	PO	PC	PO
RC	893 (TP)	129 (FN)	915 (TP)	121 (FN)	980 (TP)	252 (FN)
RO	57 (FP)	29943 (TN)	65 (FP)	29935 (TN)	142 (FP)	46051 (TN)
Acc.(C):	99.40% ± 0.21%		99.17%		99.30%	
Prec.(C):	94.23% ± 2.74%		93.37%		87.34%	
Recall(C):	87.39% ± 8.07%		88.32%		79.55%	
AUC(C):	0.931 ± 0.043		0.920		0.824	

PC: predicted covert, PO: predicted overt, RC: real covert, RO: real overt,
TP: true positive, TN: true negative, FP: false positive, FN: false negative,
(C): covert as positive class, Acc.: accuracy, Prec.: precision, AUC: area under ROC curve.

- The performance downgrade when using feature set C (high and medium relevant) if compared with the feature set A or B is considerable. Therefore, low relevant features—μ_{wS}, H_a, $p(Mo)$, $pkts$—are still determinant for the classification and should be included regardless of possible redundancies.
- The downgrade in the validation phase is obvious in spite of the used feature set. False negatives mainly belong to the novel not-trained techniques ED1 and ED2 (Table 6), meaning that new techniques have room to exploit channel capacities in ways that can bypass classifiers trained with known, old techniques.

Similar test have been carried out with other classifiers, specifically: random forests, SVM, neural networks, Bayes-based ensemble and k-Nearest

Table 6   Identification of FN and FP in the validation phase for the relevant feature set

Technique	Wrong	$\mu_{conf}(0)$	$\mu_{conf}(1)$
CAB (FN)	8	1.00±0.00	0.00±0.00
BER (FN)	18	1.00±0.00	0.00±0.00
SHA (FN)	2	0.75±0.35	0.25±0.35
GAS (FN)	24	1.00±0.00	0.00±0.00
JIN (FN)	12	1.00±0.00	0.00±0.00
LUO (FN)	8	1.00±0.00	0.00±0.00
ZAN (FN)	11	1.00±0.00	0.00±0.00
GIF (FN)	3	0.78±0.19	0.22±0.19
ED1 (FN)	99	1.00±0.00	0.00±0.00
ED2 (FN)	55	0.70±0.10	0.30±0.10
Overt (FP)	81	0.05±0.09	0.05±0.09

FP: false positive, FN: false negative, wrong: total number of wrong classified flows
$\mu_{conf}(0)$ and $\mu_{conf}(1)$ stand respectively for the average decision tree confidence to establish the 0 (overt) or a 1 (covert) label to the misclassified samples.

neighbors learners. A plain decision tree obtained the best performances[8], yet the previous discussed aspects are common in all tested learning schemes.

5.4 Unsupervised Analysis

Finally, tests by outlier ranking algorithms tried to elucidate if the new considered features could provide extra information to face the detection of covert timing channels from an unsupervised manner. Experiments in [15] revealed that covert timing channels can hardly be seen as outliers. Here, outlier analysis are performed by considering two different feature sets:

- Feature Set B: high, medium and low relevant features.
- Feature Set D: new high, medium and low relevant features not used in [15], i.e., T_R, T_S, H_a and K.

Results are shown in Table 7 and Figure 6. Results confirm that density-based outlier detection is completely useless for detecting covert timing channels. Only the HBOS algorithm (histogram-based method) can face the task to differentiate between overt and covert flows. However, the probabilities to suspect that a legitimate flow is a covert flow are still too high. For example, the P@n index gives the proportion of correct results in the top n ranks.

Table 7   Outlier detection performance indices for feature set B and D

	P@n	Adj. P@n	AP	Adj. AP	MaxF1	Adj.MaxF1	ROC/AUC
				Feature set B			
LOF	0.02	−0.01	0.01	−0.03	0.08	0.05	0.55
COF	0.01	−0.02	0.00	−0.03	0.07	0.04	0.53
INFLO	0.02	−0.01	0.01	−0.03	0.08	0.05	0.51
HBOS	0.28	0.26	0.27	0.25	0.49	0.47	0.96
				Feature set D			
	P@n	Adj. P@n	AP	Adj. AP	MaxF1	Adj.MaxF1	ROC/AUC
LOF	0.04	0.01	0.02	−0.01	0.08	0.05	0.57
COF	0.00	−0.03	0.00	−0.03	0.09	0.06	0.62
INFLO	0.03	0.00	0.02	−0.01	0.07	0.04	0.52
HBOS	0.09	0.06	0.12	0.10	0.36	0.34	0.92

---

[8]The decision tree configuration included pre- and postpruning to avoid overfitting and favor generalization. It used Information Gain (i.e., entropy-based) as splitting criterion; the minimal size for splitting was four samples; the minimal leaf size was two samples, allowing a maximal tree depth of 20 levels; the minimal gain for splitting a node was 0.1; the confidence level used for the pessimistic error calculation of pruning was 0.25, whereas the number of prepruning alternatives was three. In addition, a 10-fold cross-validation process was performed to reinforce disclosed models.

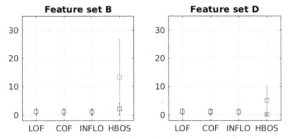

Figure 6 Outlier ranking results. Again, distributions are not normal but strongly skewed. Plots show medians and Confidence Intervals over the median (approximately 95%). Red markers correspond to covert datasets, blue for overt datasets. Red lines are indistinguishable (or almost) in LOF, COF and INFLO due to value overlap.

P@n = 0.28 in Table 7 is the best result for the HBOS case, but still too low to build a practical detector. AUC rates are nevertheless significantly high for HBOS. The AUC index can be understood as the probability that the algorithm ranks a randomly chosen positive example higher than a randomly chosen negative example. Summarizing, results show that practically all flows with covert channels obtain high HBOS ranks; overt flows are low-ranked instead, but still some overt flows—a meaningless proportion but still many when compared with covert flows rates—score high and even higher than covert cases.

Already in [15] HBOS was the only method to find a significant difference between overt and covert flows. The inclusion of the new features (T_R, T_S, H_a and K) reinforces the outlier nature of flows with covert channels, as shown in Figure 6, but is still far from being discriminant.

Therefore, the main findings of the unsupervised experiments are:

- As a general trend, covert channel flows obtain higher outlierness ranks when analyzed by the HBOS algorithm (histogram, distance-based), but still a significant proportion of overt flows scores equally high or even higher.
- The new features T_R, T_S, H_a and K endorse the outlier nature of covert channels, but the most revealing outlierness ranks are obtained when they are combined and used together with S_k, S_s, c, U, $\mu_\omega S$, $p(Mo)$ and *pkts*.

6 Conclusions

In this work we have deeply analyzed the capacity of some network traffic features to disclose covert timing channels. All analyzed features are measured or calculated from network traffic flows IATs (Inter Arrival Times).

From the analyzed features, the most relevant ones for the detection of covert channels are: c, S_k, *pkts* and K. c is the estimation of potential covert byte-equivalent symbols, which depends on the number of packets in the flow (*pkts*) and the number of multimodes that appear in the probability density estimation when using gaussians as kernels (S_k). K is the Kolmogorov complexity estimation approximated by using zlib compression. On the other hand, the hurst coefficient (H_q) and autocorrelation-based coeffcients (ρ_A) are found inefficient and can be discarded.

Conducted experiments show that the detection of covert timing channels is a demanding challenge and new covert techniques can easily bypass trained detection schemes. The high variety of forms that network traffic can take and the extremely low expected rate of covert channels in real traffic are the main reasons that make the accurate detection so difficult. Combination of supervised and unsupervised techniques (i.e., semi-supervise methods) appears as the right direction to follow in order to develop satisfactory detectors. However, a considerable rate of false positives is almost unavoidable unless more complex methods are developed and implemented.

Acknowledgment

The research leading to these results has been partially funded by the Vienna Science and Technology Fund (WWTF) through project ICT15-129, BigDAMA.

References

[1] V. Berk, A. Giani, G. Cybenko, and N. Hanover (2005). Detection of covert channel encoding in network packet delays. *Rapport technique TR536, de lUniversité de Dartmouth*, 19.

[2] James V. Bradley (1968). *Distribution-Free Statistical Tests*, 1st Edition. Prentice-Hall.

[3] M. Breunig, H.-P. Kriegel, R. T. Ng, and J. Sander (2000). Lof: Identifying density-based local outliers. *SIGMOD Rec.* 29, 93–104.

[4] S. Cabuk, C. E. Brodley, and C. Shields (2004). IP covert timing channels: Design and detection. In *Proceedings of the 11th ACM Conference on Computer and Communications Security, CCS'04*, (ACM: New York, NY, USA), 178–187.

[5] G. O. Campos, A. Zimek, J. Sander, R. J. G. B. Campello, B. Micenkov, E. Schubert, I. Assent, and M. E. Houle (2016). On the evaluation of

unsupervised outlier detection: Measures, datasets, and an empirical study. *Data Mining and Knowledge Discovery*, 30, 891–927.

[6] A. Carbone, G. Castelli, and H. E. Stanley (2004). Time-dependent hurst exponent in financial time series. Applications of Physics in Financial Analysis 4 (APFA4). *Physica A: Statistical Mechanics and Its Applications*, 344, 267–271.

[7] E. J. Castillo, X. Mountrouidou, and X. Li (2017). "Time Lord: Covert Timing Channel Implementation and Realistic Experimentation," in*Proceedings of the 2017 ACM SIGCSE Technical Symposium on Computer Science Education*, *SIGCSE'17*, (ACM: New York, NY, USA), 755–756.

[8] ElevenPaths. Low cost malware that uses gmail as a covert channel. *Telefónica Digital España*, Technical report, May 2016.

[9] W. Gasior and L. Yang (2011). "Network covert channels on the android platform," in *Proceedings of the Seventh Annual Workshop on Cyber Security and Information Intelligence Research*, *CSIIRW'11*, (ACM: New York, NY, USA), p. 61:1.

[10] J. Giffin, R. Greenstadt, P. Litwack, and R. Tibbetts (2003). *Covert Messaging through TCP Timestamps*. Springer: Berlin, Heidelberg, 194–208.

[11] M. Goldstein and A. Dengel (2012). "Histogram-based Outlier Score (HBOS): A Fast Unsupervised Anomaly Detection Algorithm," in *Advances in Artificial Intelligence*, (KI-2012), ed. S. Wölfl, pp. 59–63. [Online, 9, 2012].

[12] F. Iglesias, R. Annessi, and T. Zseby (2016). DAT detectors: uncovering TCP/IP covert channels by descriptive analytics. *Security and Communication Networks*, 9, 3011–3029.

[13] F. Iglesias, V. Bernhardt, R. Annessi, and T. Zseby (2017). "Decision tree rule induction for detecting covert timing channels in TCP/IP traffic," in *First IFIP TC 5, WG 8.4, 8.9, 12.9 International Cross-Domain Conference, CD-MAKE 2017*, in (eds) Holzinger A., Kieseberg P., Tjoa A., Weippl E., Lecture Notes in Computer Science, Vol. 10410, (Springer: Cham), 105–122.

[14] F. Iglesias and T. Zseby (2015). Analysis of network traffic features for anomaly detection. *Machine Learning*, 101, 59–84.

[15] F. Iglesias and T. Zseby (2017). "Are network covert timing channels statistical anomalies?" in *Proceedings of the 12th International Conference on Availability, Reliability and Security*, *ARES'17*, (ACM: New York, NY, USA), 81:1–81:9.

[16] W. Jin, A. K. H. Tung, J. Han, and W. Wang (2006). *Ranking Outliers Using Symmetric Neighborhood Relationship*. Springer: Berlin Heidelberg, 577–593.

[17] T. Kleinow (2002). *Testing continuous time models in financial markets.* PhD thesis, Humboldt-Universitát zu Berlin, Wirtschaftswissenschaftliche Fakultt, 2002.

[18] A. N. Kolmogorov (1968). Three approaches to the quantitative definition of information. *International Journal of Computer Mathematics*, 2, 157–168.

[19] M. Li and P. Vitanyi (1997). *An Introduction to Kolmogorov Complexity and Its Applications (Texts in Computer Science)*. Springer.

[20] X. Luo, E. W. W. Chan, and R. K. C. Chang (2008). "TCP covert timing channels: Design and detection," in *IEEE International Conference on Dependable Systems and Networks with FTCS and DCC (DSN)*, 420–429.

[21] Henry B. Mann (1945). On a test for randomness based on signs of differences. *Ann. Math. Statist.* 16, 193–199.

[22] N. Meinshausen and P. Bhlmann (2010). Stability selection. *Journal of the Royal Statistical Society: Series B (Statistical Methodology)*, 72, 417–473.

[23] B. H. Menze, B. M. Kelm, R. Masuch, U. Himmelreich, P. Bachert, W. Petrich, and and F. A. Hamprecht (2009). A comparison of random forest and its Gini importance with standard chemometric methods for the feature selection and classification of spectral data. *BMC Bioinformatics*, 10, 213.

[24] MEJ Newman (2005). Power laws, Pareto distributions and Zipf's law. *Contemporary Physics*, 46, 323–351.

[25] M. A. Padlipsky, D. W. Snow, and P. A. Karger (1978). Limitations of end-to-end encryption in secure computer networks, 1978. ESD-TR-78-158.

[26] H. Peng, F. Long, and C. Ding (2005). Feature selection based on mutual in- formation criteria of max-dependency, max-relevance, and min-redundancy. *IEEE Transactions on Pattern Analysis and Machine Intelligence*, 27, 1226–1238.

[27] S. M. Pincus (1991). Approximate entropy as a measure of system complexity. *Proceedings of the National Academy of Sciences*, 88, 2297–2301.

[28] G. Shah, A. Molina, and M. Blaze (2006). "Keyboards and covert channels," in *Proceedings of the 15th Conference on USENIX Security Symposium, USENIX-SS'06*, USENIX Association: Berkeley, CA, USA.

[29] B. W. Silverman. Using kernel density estimates to investigate multimodality. *Journal of the Royal Statistical Society: Series B*, 43, 97–99.

[30] J. Tang, Z. Chen, A. W. Fu, and D. W.-L. Cheung (2002). "Enhancing effectiveness of outlier detections for low density patterns," in *Proceedings of the 6th Pacific-Asia Conference on Advances in Knowledge Discovery and Data Mining, PAKDD'02*, (Springer-Verlag: London, UK), 535–548.

[31] TU Wien CN Group. Data Analysis and Algorithms, 2017.

[32] J. Wu, Y. Wang, L. Ding, and X. Liao (2012). Improving performance of network covert timing channel through Huffman coding. *Mathematical and Computer Modelling*, 55, 69–79, 2012.

[33] S. Zander, G. Armitage, and P. Branch (2007). "An empirical evaluation of ip time to live covert channels," in *2007 15th IEEE International Conference on Networks, ICON 2007*, (IEEE), 42–47.

Biographies

Félix Iglesias Vázquez was born in Madrid, Spain, in 1980. He obtained the Dipl.-Ing. in electrical engineering and MAS in IT from the Ramon Llull University, Barcelona, Spain. In 2012 he received the Ph.D. degree in technical sciences from TU Wien, Austria, where he currently holds a University Assistant position doing fundamental research in data analysis and network security. He has worked on R&D for diverse Spanish and Austrian firms, and lectured in the fields of electronics, physics, automation, machine learning and data analysis.

Robert Annessi received his B.Sc. and his M.Sc. degrees in computer engineering from TU Wien in 2011 and 2014 respectively. He is genuinely interested in communication networks, network security, and privacy, and is currently pursuing his Ph.D in the area of secure group communication for critical infrastructures. His further research interests are anonymous communication, covert communication, and subliminal communication.

Tanja Zseby is a professor of communication networks in the Faculty of Electrical Engineering and Information Technology at TU Wien. She received her Dipl.-Ing. degree in electrical engineering and her Ph.D. (Dr.-Ing.) from Technical University Berlin, Germany. Before joining TU Wien she led the Competence Center for Network Research at the Fraunhofer Institute for Open Communication Systems (FOKUS) in Berlin and worked as visiting scientist at the University of California, San Diego.

An Approach for Building Security Resilience in AUTOSAR Based Safety Critical Systems

Ahmad MK Nasser[1], Di Ma[1] and Priya Muralidharan[2]

[1] University of Michigan, Dearborn, USA
[2] Renesas Electronics America
E-mail: {ahmadnas,dmadma}@umich.edu; priya.muralidharan@renesas.com

Received 5 December 2017; Accepted 6 December 2017;
Publication 19 December 2017

Abstract

AUTOSAR, a worldwide development partnership among automotive parties to establish an open and standardized software architecture for electronic control units (ECUs), has seen great success in recent years by being widely adopted in deeply embedded automotive ECUs. Increasing the security resilience of AUTOSAR based systems is a crucial step in securing safety critical automotive systems. We study AUTOSAR safety mechanisms and demonstrate how they can be used as attack vectors to degrade the vehicle safety. We show the need to harmonize the fail-safe response with the secure state of the system. And we evaluate the overlap in the properties of safety mechanisms with security objectives to highlight methods for hardening automotive systems security.

Keywords: Automotive cyber security, AUTOSAR, Safety, Security, ISO26262, Embedded security.

1 Introduction

Protecting the vehicle control side against cyber attacks is the ultimate goal of any robust security architecture. With added connectivity, even deeply embedded devices are under the threat of cyber attacks [23, 24, 28].

Journal of Cyber Security, Vol. 6_3, 271–304.
doi: 10.13052/jcsm2245-1439.633

The safety standard for automotive systems, ISO26262 [26], defines a formal approach for the creation of safety mechanisms to mitigate systematic and random faults in the system. However, malicious attacks are not considered by the failure response of the system. When a safety relevant failure condition is detected, the system switches to a safe state, also known as a fail-safe state. Depending on the risk level associated with the related hazard, a safe state may range from disabling a control function (e.g. park assist in the case of a sensor failure), to resetting the system (e.g. in the case of detected memory corruption). Safety mechanisms which are adequate in the face of normal failures, are inadequate and in some cases, even themselves exploitable, in the presence of a malicious attacker [25]. Injecting a safety fault can thus launch the equivalent of DoS attack against the vehicle safety functions. While there is a real need for hardening the security of deeply embedded systems, it has to be done within the cost and performance constraints of automotive systems.

The need for security-in-depth measures for securing automotive systems is well-understood. At the ECU level, this means participating in secure communication, supporting for code signing of flash downloads, securing the diagnostic interface, securing the debug interfaces, and using secure boot. Even as the automotive industry takes concrete steps to apply such best practices, software bugs are still expected to present major challenges to comprehensive vehicle security. For security, AUTOSAR defines a cryptographic security stack to provide fundamental security services. It also defines safety mechanisms with overlapping security properties. The goal of this research is to study AUTOSAR safety mechanisms as potential sources of attacks. Also to answer the question of how shall a system differentiate a safety failure which happens due to a deterministic fault vs. a failure injected through an attack in order to harmonize the safe and secure response. Moreover, is it possible to leverage existing AUTOSAR safety mechanisms with security properties to increase the system security? By studying this area, we aim to improve the security resilience of AUTOSAR based systems within the constraints of the automotive environment.

Toward this goal, we make the following contributions: we take a systematic approach to evaluate safety mechanisms that could be potentially exploited, we demonstrate two such attacks on a safety qualified target, and then propose corresponding countermeasures that can be applied for similar types of attacks. We also experiment with using the AUTOSAR OS stack interference protection as a security mechanism to prevent a buffer overflow attack. We show how AUTOSAR safety mechanisms can be used as security mechanisms and we define the constraints to enable such use. We also demonstrate scenarios in which the security response in the presence of an attack can violate the security response and vice versa to highlight the need for a harmonized process for fail safe and secure action.

The organization of this paper is as follows. In Section 2, we briefly discuss related work. In Section 3, we present the attack model considered for our analysis. In Section 4, we offer a survey of AUTOSAR safety features. In Section 5, we present a method for assessing safety mechanisms as either exploitable or dually useful for security protection. In Section 6, we analyze safety mechanisms as exploitable candidates. In Section 7, we analyze safety mechanisms that have overlapping security properties. In Section 8, we present the results of our attacks on safety mechanisms, while in Section 9 we demonstrate the efficacy of a safety mechanism as a security mechanism. In Section 10 is the conclusion and future work.

2 Related Work

The use of an automotive error mitigation mechanism as a security attack vector was studied in [13]. The authors demonstrated an attack that uses the CAN busoff error response to stop an ECU from communicating. The Bus-Off condition is designed to prevent an ECU from disturbing the rest of the bus by forcing it to go off line after a certain number of transmission errors is detected. By creating message collisions with a target CAN frame an attacker could induce a Bus-Off error condition in the target ECU. This is similar to an exploitable safety mechanism but we take a systematic approach to finding such exploits and we do it by studying AUTOSAR safety mechanisms.

The conflicts and overlap between safety and security for automotive systems was studied in [21]. The work highlighted the need for a holistic approach to designing both safe and secure systems. One example conflict area is the cyclic RAM test that requires stopping CPU cores except the one performing the test in order to prevent a false detection of RAM corruption. In systems where a hardware security module (HSM) is active, stopping the HSM violates a security principle. Not stopping the HSM can result in corruption of the shared memory accessible by the HSM and the other CPUs. This points to an area of conflict that should be addressed at the system level. The paper also discusses areas of overlap between safety and security such as using the memory protection unit (MPU) to provide both freedom from interference as well as security isolation. Our goal in this paper is to increase the resilience of AUTOSAR systems by eliminating exploits and adding defenses strictly from the perspective of safety.

Attacks against embedded systems were studied in [18]. The authors demonstrated control flow and stack overflow attacks on embedded systems and then proposed a hardware based solution for the prevention of such attacks. Their attacks were relevant to our work as we aim to provide defense mechanisms for embedded systems against the types of attacks presented in their work. The use of the CPU dual lock step safety mechanism as a security countermeasures was studied in [30]. The authors investigated the resilience of ASIL-D microcontrollers to power glitching attacks. They showed that the dual lock step mechanism failed to detect glitch attacks allowing them to jump over instructions and thus compromising the security of an ECU. The paper shined a light on how safety mechanisms needed to evolve to protect against security attacks. Our research takes this concept a step further by performing a comprehensive survey of AUTOSAR safety mechanisms to find areas in which safety mechanisms could be useful to increase the security of safety critical systems.

3 Attack Model

Deeply embedded systems are assumed to be located behind several defense lines: firewall, security gateway, and a secure communication bus. Assuming the vehicle has implemented a properly layered secure architecture, launching successful attacks on such systems requires compromising several security layers upstream. Previous works by [24] and [28], have shown that at some point these defenses can be broken and an attacker may be able to launch a successful attack on the vehicle control system. Let us consider the three classes of attacks that are relevant to both deeply embedded ECUs and traditional computer systems [15]:

1. Malware or exploitable software vulnerabilities to take control of the system
2. Physical access type attacks
3. Network based attacks

Note in the first class, there is a significant difference in the attack surface between traditional computers and deeply embedded systems [12]. In the former, exposure to software exploits is more common due to the rich space of applications that can be loaded and executed on highly configurable operating systems like Linux. In contrast, deeply embedded systems execute a limited pre-defined set of applications from flash memory. Creating persistent malware in flash, requires the tampering of the flash bootloader or an exploit that can bypass security checks to use the bootloader routines directly.

On one hand, loading temporary malware requires injecting code in data memory (such as the stack) through a buffer overflow exploit due to a software bug as demonstrated in [17]. When network access is considered together with software based exploits, deeply embedded systems become targets of similar types of attacks as traditional computer systems. While security experts may argue that many protections are already being designed to harden automotive systems, lack of maturity of cyber security principles within Automotive ECUs gives us the intuition that software vulnerabilities and back doors will persist for several years. On the other hand, in case of ADAS systems which are expected to provide a certain level of autonomy, such systems may no longer exist in a deeply embedded layer, but rather in a mixed criticality system on rich execution environments which are closer to the attackers access than say a brake controller. For such systems, the safety functions are expected to run in a partitioned execution area that is supposedly isolated from the rest of the system. That is the focus of AUTOSAR Adaptive Platform [19] and is out of scope for our current analysis.

For the rest of this paper we assume our attackers have direct network access to the ECU, the network supports CAN authentication, and there exists an exploit in the ECU which allows loading software on the target. We also assume that the ECU has a secure element (HSM) which serves as the root of trust for the system. As for the attacker's objective, it is to disrupt safety critical systems for the aim of inflicting harm on drivers, damaging an OEM reputation, or creating ransom-ware that locks up safety functions.

4 Survey of AUTOSAR Safety Features

4.1 End to End Library

The first AUTOSAR module that we introduce here is the End to End (E2E) library which defines several protection profiles for data transmitted over a communication channel both internally and externally [20]. The goal of this module is to prevent safety critical functions from operating on faulty or missing data as shown below:

1. CRC check to detect corruption of data
2. Sequence counter to detect out of sequence messages
3. Alive counter to prevent operating on old data
4. A unique ID for Interaction Layer Protocol Data Unit (I-PDU) group to detect a fault of sending I-PDU on unintended message
5. Timeout monitoring to detect communication loss with the sender

Mechanisms 1 through 4 listed above allow the detection of non-malicious errors in the content of a message. The timeout detection mechanism protects an ECU from operating on old data due to loss of communication with the data source. To achieve this, a message is monitored by the receiving node based on its expected periodicity. If the message is not received by the expected deadline, AUTOSAR provides a mechanism to log a timeout failure and take fail safe action such as disabling the consuming function.

4.2 AUTOSAR OS

AUTOSAR OS [7] is a real time operating system specification. The standard defines protection mechanisms that are essential for building safety critical applications. Those protections fall under the following categories:

1. Memory Protection: to provide freedom of interference between OS applications and tasks of mixed criticality.
2. Timing Protection: to prevent timing errors in tasks, Interrupt Service Routines (ISRs), or system resource locks from interfering with higher ASIL functions.
3. Service Protection: to capture invalid use of the OS services by the application.
4. OS Related Hardware Protection: to protect privileged hardware elements from being modified by lower ASIL functions.

Note the OS supports four scalability classes with the following features:

1. SC1: Deterministic Real time operating system (OSEK OS based)
2. SC2: Stack monitoring and precise time control for periodic tasks
3. SC3: Support for MPU/MMU to provide spatial freedom of interference
4. SC4: Timing protections

4.2.1 Memory protections

AUTOSAR OS SC3 and SC4 support freedom of interference between software partitions of mixed safety criticality through hardware based spatial separation of memory [7]. The aim of these protections is to prevent a lower ASIL software from corrupting the data of a higher ASIL software within the same system. To understand the hardware based memory protection capabilities of automotive embedded systems, we studied the ARM Cortex M architecture which is among the most popular micro-controller architectures used in automotive ECUs [11]. Rather than an MMU, such MCUs rely on an MPU to achieve the desired goal of freedom from interference. Also the CPU supports two modes: privileged and user mode. Only privileged mode allows access to special registers like the MPU configuration. Using the MPU it is possible to define memory regions with specific attributes such as read, write, and execute as well as specify access rights by privileged or user modes. AUTOSAR OS supports protecting memory both at the Task/ISR Cat2 level and at the OS application level. When switching to a "non-trusted" OS application, the OS reconfigures the MPU to restrict access to the Safety Application code, data and private stack. This prevents a lower ASIL OS application from being able to corrupt the data of a higher ASIL OS application.

Note in addition to MPU based stack protection, AUTOSAR OS defines software based stack monitoring which can only identify where a task or ISR has exceeded a specified stack usage at context switch time. This is done by checking a unique stack pattern which is inserted at the end of the reserved stack space. This can result in considerable time between the system being in error and the fault being detected. Besides data protection, the MPU can be used to restrict access to memory mapped registers to prevent certain tasks from modifying hardware registers which are safety relevant. Note, AUTOSAR uses the term trust in the context of safety which can be misleading because Cyber Security threats are not considered. Consequently, it is possible to define a "Trusted OS Application" that has access to all memory resources even though from a security point of view, that OS Application may be vulnerable to attacks.

4.2.2 Timing protections

AUTOSAR OS timing protections aim to mitigate timing faults that can exist in lower ASIL software from propagating to higher ASIL software. The timing protections are:

1. Execution Time Protection: detects faults in Tasks or Category 2 ISRs that exceed their execution budget.
2. Locking Time Protection: detects faults in blocking resources, and locking interrupts for a period longer than the configured maximum.
3. Inter-arrival time protection: detects faults in the time between successive activations of tasks or Cat2 interrupts to ensure a minimum time is not violated.

By monitoring the execution time of tasks and Category 2 ISRs, AUTOSAR OS aims to detect timing errors before they can lead to tasks missing their deadlines. This prevents the propagation of timing errors to higher priority tasks and allows the OS to isolate the offending task. Note the OsTaskExecutionBudget is a configurable parameter that specifies the maximum allowed execution time of a task [7]. One use case for the locking time protection mechanism is to prevent global interrupts from being disabled for a period of time that would create instability in the real time system. Disabling global interrupts is needed in scenarios where a routine needs atomic access to a resource and cannot allow being interrupted. But doing so beyond a specific time threshold can prevent the real time system from being able to process critical tasks within the required time. The third timing protection mechanism is needed to ensure the system is not being excessively interrupted or activating tasks.

4.3 Hardware Protection

AUTOSAR OS is expected to run in privileged mode which gives it access to special hardware registers and protects those registers from corruption by Tasks or Cat2 ISRs running in user mode. Example registers that are only accessible in supervisor mode can be the MPU configuration, the OS Timer unit, and the interrupt control configuration. Protecting those registers from faults in lower ASIL software is mandatory to ensure the integrity of the OS operation.

4.4 Watchdog Manager

AUTOSAR defines three modules for supporting watchdog functions [9]:
1. Watchdog Driver: services the hardware watchdog whether internal or external
2. Watchdog Interface: provides a high level of abstraction of watchdog driver functions
3. Watchdog Manager: supports the supervision of multiple software entities and the triggering of an MCU reset in case of a supervision failure

Supervised entities can be software components, runnables, or Basic Software (BSW) modules and they are marked by checkpoints. The user configures checkpoints based on code sections which are deemed to be critical for the safety of the system. Using such checkpoints it is possible to monitor aliveness, timeout, and control flow within a supervised entity.

The user inserts calls to WdgM_CheckpointReached() in locations where the WdgM shall be notified of an execution event of a supervised entity.

A software error that prevents the checkpoint from being reached by the deadline or with the right order results in the detection of a fault by the WdgM during the execution of the WdgM_Mainfunction. The response to any such fault can range from a simple notification to an MCU reset.

4.5 Core Test

Safety critical applications require the monitoring of an MCU core functions to detect hardware faults during startup or normal runtime of an ECU. The core test module [2], can perform tests of MCU components such as:

1. Arithmetic Logic Unit (ALU)
2. Memory Protection Unit (MPU)
3. Cache controller
4. Interrupt Controller

The core test is executed in partial tests as a background task that can be interrupted by higher priority tasks. However, the core test requires uninterrupted execution of atomic sequences. In case a core test fails, the module reports the event to the Diagnostic Event Manager (DEM) [4] to take action based on the severity of the detected failure. Note that microcontrollers with dual lock step cores do not need the Core Test module since the dual core lock step feature can detect errors covered by this module.

4.6 RAM Test

The RAM test module [8] provides a physical health test of RAM cells and RAM registers to meet the fault coverage requirements of a safety critical application. The tests can either be executed in a background task or through a direct call from the application. In case a failure is detected, the module reports the results to the DEM to take the appropriate action. AUTOSAR defines interfaces to start and stop the tests but there is no direct interface to set the test status to a failure. The module splits tests into atomic units that are not interruptible. A higher priority task can interrupt the module test in between atomic test units execution. The background task performs the tests of all configured blocks sequentially and repeats the sequence after each complete test is finished. One of the limitations of this module is that during the execution of the RAM test algorithm, another software shall not attempt to modify the area under test. This is to ensure data consistency in multi-core systems or with DMA controllers. The AUTOSAR specification lists the following algorithm types that are supported:

1. Checkerboard test algorithm
2. March test algorithm
3. Walk path test algorithm
4. Galpat test algorithm
5. Transparent Galpat test algorithm
6. Abraham test algorithm

4.7 Flash Test

The Flash Test module [5], provides test algorithms for non-volatile memory to meet the diagnostic coverage requirements in a safety critical system. Tests are divided into partial tests based on the number of cells tested in one task cycle. Unlike RAM test and Core Test, the Flash test can be pre-empted at any point because it does not require atomic access. It is possible to abort or suspend the flash test but that introduces a latency based on when the request is received related to the background task cycle time. A failure during the test algorithm is reported to DEM to take the proper action. The different test algorithms supported are:

1. 16 Bit CRC
2. 32 Bit CRC
3. 8 Bit CRC
4. Checksum
5. Duplicated Memory
6. ECC

5 Safety Mechanism Classification Method

We propose a method for classifying AUTOSAR safety mechanisms as either candidates for: security exploits, security protections, or neither. In Figure 1, we present the process of finding an exploit candidate. Note, here the attacker's intent is to inject a failure that mimics a safety failure in order to shutdown the safety function. First we check if the mechanism can be triggered through a network based attack. Intuitively, an attacker that has already established a foothold in the ECU has very little incentive to trigger the error response of a safety mechanism. Arguably, once an attacker is inside the ECU, his goal is to evade any protection mechanisms which is the opposite of triggering a safety mechanism. Next we check if the configured error response of the safety mechanism results in disabling a safety function or more (such as a system reset). If so, we classify this mechanism as an exploitation candidate. Here

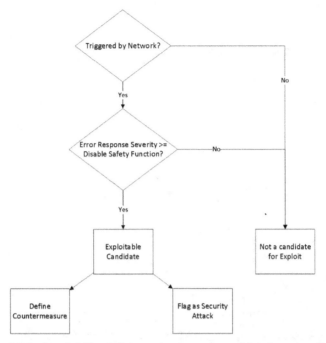

Figure 1 Method to classify a safety mechanism as a candidate for a security exploit.

we define two possible routes: define a security countermeasure to prevent the attack, and for cases where prevention is not possible, extend the fail-safe response to capture additional indicators that can help in flagging this event as a security attack.

In Figure 2, we present the process for classifying a safety mechanism as a dual purpose security mechanism. First we check if the mechanism is able to detect an attack. This can be an internal attack through malware, a physical attack like power glitching, or a network attack. Safety mechanisms that exhibit properties to allow memory isolation, hardware resource protection, control flow protection and hardware tampering detection are all example candidates for use as security mechanisms. But having a security property is not sufficient if the mechanism can be easily disabled by an attacker. Therefore we define a set of security constraints to establish trust in the usage of the safety mechanism as an attack prevention or detection mechanism. The third step is to ensure that the fail-safe response is harmonized with the required security response and vice versa. For example, in case of a mechanism that detects the presence of a malware in RAM, the security response would be to immediately

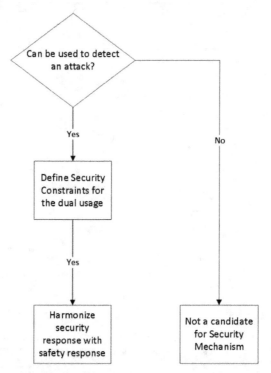

Figure 2   Method to classify a safety mechanism as a dual purpose security mechanism.

issue a reset and force a secure boot check to restore the system to a known trusted state. But issuing an immediate reset may violate a safety goal that does not allow the sudden loss of a critical safety function, for e.g. power steering. Therefore, an activity of harmonizing the safe and secure response is needed between the safety and security experts during the design of the system.

In the following sections, we apply these two methods to classify our surveyed safety features.

6 Exploitable Safety Mechanisms

This section was based on our work in [25]. There we evaluated safety mechanisms in the AUTOSAR E2E library, and AUTOSAR OS timing protections as security exploit candidates. With the E2E library, our aim was to induce a failure that would trigger a protection mechanism which would result in disabling safety critical functionality in an ECU. With AUTOSAR

OS timing protections, our aim was to trigger a system shutdown by causing a software task to exceed its runtime budget.

6.1 Attacks on E2E Protection

We start by applying the classification flow in Figure 1 to the safety mechanisms in Table 1. We assume that the network supports message authentication which can protect against spoofing attacks. This prevents an attacker from tampering with the alive counter, CRC, sequence counter or I-PDU Id protection mechanism without detection because they are covered by the message authentication code(MAC). Messages that contain invalid MAC are ignored by the receiver and therefore do not result in a trigger of a safety mechanism. What remains then is the timeout protection mechanism which can be triggered externally even when message authentication is enabled. In order to create a message timeout event, the attacker can flood the bus with high priority CAN messages, e.g. zero-ID CAN messages, at the highest periodicity possible for the target baud rate.

Table 1   Survey of AUTOSAR safety features

Module	Mechanism	Max Error Response
E2E	CRC:data integrity	Disable consuming function
E2E	Sequence counter: message order	Disable consuming function
E2E	Alive counter: data freshness	Disable consuming function
E2E	Data Id: detect I-PDU sent on wrong message	Disable consuming function
E2E	Timeout monitoring: detect message loss	Disable consuming function
WdgM	Monitor aliveness of supervised entities	MCU reset
WdgM	Detect timeout of supervised entity	MCU resets
WdgM	Monitor control flow of supervised entity	MCU resets
OSTiming	Monitor task/ISR execution budget	OS Shutdown
OSTiming	Monitor task/ISR inter-arrival time	OS Shutdown
OSTiming	Locking time protection	OS Shutdown
OSMemory	Stackoverflow detection	Reset
OSMemory	Detect execution from data section	Reset
OSMemory	Detect access to restricted memory	Reset
OSHardware	Prevent untrusted apps from accessing privileged HW	Reset
CoreTest	Test health of core MCU components	Reset
RAMTest	Test health of RAM cells	Reset
FlashTest	Test health of Non-volatile memory	Reset

Transmitting zero-ID frames in a back to back fashion will reduce the likelihood that a valid frame wins arbitration to be transmitted on the bus. As a result of transmitting nodes continuously losing arbitration to the zero-ID message, receiving nodes will start logging timeout faults. Subsequently, control functions that rely on those messages will be degraded, which is the safe state of missing safety critical messages. An attacker determined to prevent the safety critical ECU from performing its intended function can successfully launch this attack by exploiting this mechanism.

6.1.1 Countermeasures

In [25] we showed several countermeasures:

1. Add a smart gateway that runs intrusion detection software to monitor the received CAN message identifiers along with their expected frequency to detect attacks such as the zero-ID flood attack.
2. Add local secure monitor software within transmitting ECU's to detect the malicious manipulation of the CAN configuration. The monitor can either reside in the HSM or be monitored periodically by the HSM to ensure it is not disabled. If the CAN settings are flagged as tampered with, the HSM can reset the micro to prevent the disturbance of the local network.

Since this attack is not fully preventable, it is necessary to collect indicators when timeout events occur in order to distinguish normal failures from security attacks. This can be achieved by logging the frequency of these failures, as well as capturing additional network traffic to aid in the anomaly detection either through a local or off-board intrusion detection system. Although the zero-ID attack has already been mentioned in other publications, such as [24], the attack is still worth mentioning here because we arrive at it by considering the safety protection mechanisms, rather than by pure brainstorming techniques. We also stress the need to enhance the logging of such an event to identify potential malicious root causes.

6.2 Task Execution Time Attack

By applying the classification flow in Figure 1 to the safety mechanisms in Table 1 we arrive at our second exploit candidate: Task Execution Budget

Monitoring. In response to this type of timing error, the OS defines the following possible actions that the application can request [7]:

1. PRO_IGNORE: the OS can ignore the event
2. PRO_TERMINATE_TASKISR: the OS shall forcibly terminate the task
3. PRO_TERMINATE_APPL: the OS shall terminate the faulty OS Application
4. PRO_TERMINATE_APPL_RESTART: the OS shall terminate and then restart the faulty OS Application
5. PRO_SHUTDOWN: the OS shall shutdown itself

The last action from the above list implies that the system can be completely shutdown as a result of such an error condition. In a stable system absent from a malicious attacker, such a fault is normally caught during development when the system is tested under maximum load conditions. AUTOSAR OS gives the system configurator the flexibility to specify the appropriate value for the execution budget as well as the proper behavior in case it is exceeded. In the presence of an attacker who is able to repeatedly cause this error condition, the system can experience constant resets that prevent it from ever being able to execute normally.

Upon detecting the error condition the OS triggers a ProtectionHook to notify the application to take fail safe measures. As long as the application does not ignore this condition, the second condition of Figure 1 is satisfied. For the rest of this section we will see how this fault can be triggered externally.

As mentioned in Section 4.2.2, AUTOSAR OS monitors the task execution time, to prevent a single task from starving the CPU from runtime resources. To exploit this safety mechanism, the attackers goal is to cause an OS task to exceed its execution budget. One way to find candidates for this type of exploit is scanning the application for processes that have variable execution time due to their dependence on a hardware resource like flash programming time, or a network resource like processing CAN data.

We chose the latter and we investigated CAN networks that support authenticated messages as proposed in AUTOSAR 4.2 via the Secure On Board Communication (SecOC) module [6]. SecOC is a software module that provides secure on board communication support. When a secure Protocol Data Unit (PDU) is received, SecOC receives an indication from the Protocol Data Unit Router (PDUR) module to copy the PDU to its own memory buffers. It then triggers the verification of the authenticator portion of the PDU by calling the AUTOSAR Cryptographic Service Manager (CSM) module as illustrated in Figure 3. Only if the verification passes, SecOC then notifies the

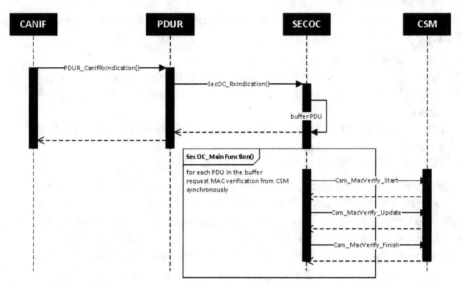

Figure 3 Sequence diagram showing data flow from CANIF to the CSM layer for frame authentication.

PDUR module to route the PDU up to the consuming layers [6]. Since SecOC relies on the SecOC_MainFunction() to perform the verification processing, the attack goal is to cause that function to exceed the AUTOSAR configured runtime budget: OsTaskExecutionBudget.

Based on the CAN FD specification [27], we can estimate the nominal time for transmitting a CAN frame if the arbitration rate, data rate and payload size are all known. As shown in Figure 8, the number of bits in a CAN FD frame can be calculated based on the different segments of the frame. Note, the length of the CRC field is either 17 bits or 21 bits depending on the payload size. For simplification, we set the CRC field to be 21 bits which corresponds to a payload length of 20 and 64 bytes. This choice is guided by the fact that in a vehicle CAN FD frames are more likely to utilize the larger payload size. Thus the only unknown variable parameter remaining is the number of stuff bits which depends on the content of the CAN frame. The rule is that no more than 5 bits can be transmitted consecutively with the same polarity. Therefore, stuff bits are inserted to ensure bit polarity is toggled if more than 5 consecutive bits have the same logic level.

Accounting for all the variables, results in a formula that gives us the estimated transmission time of a CAN FD frame (in seconds):

$$Tcanfd = (1 + f) * (30/a + (28 + dl * 8)/d). \tag{1}$$

where a is the arbitration baud rate in bits per seconds, f is the stuff bit factor, d is the data baud rate in bits per seconds, and dl is the frame data length in bytes. Note that in a worst case scenario 1 stuff bit is inserted for every 5 consecutive bits which is equivalent to a factor of 20%.

In the case that CAN message authentication is enabled the attacker takes advantage of the fact that an ECU needs to spend a fixed amount of CPU runtime to perform a MAC authentication before the frame is accepted or discarded. Note, an attacker does not have to worry about generating valid MAC values, because the goal is to exploit the time taken to verify the MAC, not to spoof a message with a valid MAC. The processing time varies depending on the target micro-controller and the CPU operating clock frequency. SecOC defines a parameter for the number of authenticating attempts when the freshness counter is not transmitted in its entirety within the frame. The parameter:SecOCFreshnessCounterSyncAttempts, causes the re-authentication of a secured I-PDU with different freshness values within the acceptance window until one authentication succeeds or all attempts fail. This results in more processing time for each message authentication failure. Therefore, this parameter shall be accounted for in the attack potential evaluation. As shown in Figure 3, SecOC_MainFunction() loops through all the buffered PDUs that require verification and triggers the verification request to the AUTOSAR CSM [3] module. We intentionally choose to configure CSM to run in synchronous mode so as to maximize the processing time spent in SecOC_MainFunction as it tries to authenticate all frames in the buffer before the task is finished.

As a result, SecOc_MainFunction() has to wait for the three CSM steps to be completed before it starts processing the next secure PDU. To achieve a successful attack, the attacker needs to send a burst of authenticated PDUs that would result in the SecOC_MainFunction() exceeding its runtime execution budget. The key here is finding the minimum size of the frame burst needed to cause the timing error condition and then checking whether it is feasible given the constraints of the CAN FD protocol. The attack is possible if there exists a value $B \leq maxB$ such that $T_{processing} > T_{budget}$ where:

$$maxB = \frac{T_{secoc}}{T_{canfd}}. \tag{2}$$

Therefore, assuming SecOC_MainFunction has a task cycle time of T_{secoc} and a CAN FD frame transmission time of T_{canfd} our goal is to find the minimum burst size B such that the processing time of SecOC_MainFunction $T_{processing}$ is greater than the configured execution budget T_{budget} while $B \leq maxB$.

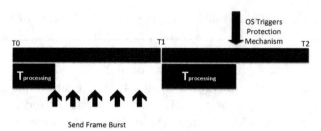

Figure 4   Triggering the OS protection mechanism by causing the SecOC main function task to exceed the execution budget.

In order to evaluate if a system is affected by this attack, we present Equation (3) for calculating burst size B. Let T_{mac} be the MAC verification time for verifying a single 64 byte message, note this time depends on the MAC algorithm and whether it is accelerated in hardware or implemented in software. Let T_{main} be the runtime to execute the SecOC_MainFunction() to process a single frame without the MAC calculation overhead. Let T_{budget} be the maximum execution budget of the SecOC_MainFunction task. Let $N_{attempts}$ be the value of SecOCFreshnessCounterSyncAttempts, which is the number of attempts performed if MAC verification fails. Let B be the number of CAN FD messages that can be verified within the T_{budget} time:

$$B = \frac{T_{budget}}{(N_{attempts} * (T_{main} + T_{mac}))} \tag{3}$$

The above analysis gives us the conditions needed to determine if the mechanism can be triggered externally based on the target system parameters. An evaluation on a real target is shown in Section 8.2.

6.2.1 Countermeasures
The countermeasures to this attack as shown in [25] consist of the following:

1. When defining Task Execution Budgets, system designers shall take into account potential security threats to the task execution time to find the optimal budget that can address both safety and security related faults.
2. Wherever possible, choose the asynchronous mode for performing specific functions. AUTOSAR provides both synchronous and asynchronous mode in several modules like NVM and CSM. This would limit the time spent in a task as it allows the job to be performed over several call cycles.
3. With CAN authentication, SecOC defines a maximumretryCounter to repeat the MAC verification with a different freshness counter upon

failure. This can further exacerbate the duration of performing the CAN authentication when an attacker sends a burst of invalid messages (wrong MAC). Therefore, using the minimum value possible for this parameter is recommended.

4. Use a network anomaly detection system that can flag and stop anomalous message bursts like the one used in this paper to trigger the attack.
5. A fatal error such as a system shutdown due to exceeding the execution budget may seem highly improbable under normal conditions, but under the influence of an attacker can become much more likely. Therefore, it is necessary to review all fatal errors in the application to re-evaluate if the conditions of such errors are impacted by a malicious attacker.

In the cases where the attack cannot be prevented, it is necessary that the fail-safe response collects additional indicators to aid in future forensic analysis. Some of those indicators can be the network traffic at the time the fault occurred, as well as the frequency at which the error condition occurred. This information can be analyzed by a vehicle IDS or an offline system that can observe inputs from many vehicles to identify fleet wide attacks.

7 Dual Purpose Safety Mechanisms

Being able to use AUTOSAR safety mechanisms for attack detection can increase the security resilience of AUTOSAR based systems without a major impact to the cost of the system. In the below sections we show the results of our analysis of AUTOSAR safety mechanisms as means for attack detection.

7.1 Analysis

7.1.1 Stack usage monitoring

Code injection attacks through buffer overflow [16] continue to be among the most effective in computer systems, where an attacker can overflow a buffer boundary in the task stack in order to overwrite the return address of a function. This results in rerouting the flow of the program to the new address overwritten by the overflow.

Naturally, a protection mechanism that can detect stack overflow would be relevant for security. As mentioned in Section 4.2.1, AUTOSAR OS supports memory stack overflow monitoring either through software checks (by checking special patterns on the stack) or with the help of the MPU. Since a malicious attacker can easily forge the stack pattern value after overwriting the stack space, software based stack protection cannot be considered a security mechanism. With the MPU based stack protection, the OS sets up a dedicated MPU stack entry prior to activating the corresponding task. The user defines the stack size for each context based on prior measurements to determine the maximum required stack size. An attacker who manages to inject code in a stack space cannot exceed the stack boundary and cannot inject code in a stack dedicated for another context. Doing so results in an immediate exception which results in the OS taking corrective action such as issuing a reset. This matches well as a mechanism for attack detection.

7.1.2 RAM execution prevention

Modern operating systems support data execution prevention (DEP) to ensure that a memory address can either be configured as writable or executable but not both. Automotive embedded operating systems do not explicitly support such protections, but using AUTOSAR OS, it is possible to emulate this protection. As mentioned in Section 4.2.1 using an MPU it is possible to setup access rights to memory regions as: read, write, and execute. While AUTOSAR OS does not explicitly define how to separate access rights, it is expected that specific vendor implementations of AUTOSAR OS offer the user the ability to assign different access rights to different OS applications.

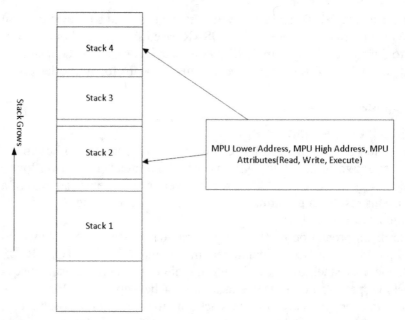

Figure 5 MPU based stack protection, by switching MPU setting based on the active stack.

We propose extending this to restrict all execution out of RAM regardless of the safety level of the corresponding application. While this does not prevent all stack based attacks as is the case with a return-to-libc attack [14], it does raise the difficulty level for mounting attacks on embedded systems. An attacker who attempts to violate this rule will immediately cause a CPU exception which can reset the system to restore it to a known secure state.

7.1.3 Control flow integrity protection

The topic of control flow integrity(CFI) was thoroughly studied in [10]. Moreover, [17] showed several techniques to alter the execution flow specifically in an embedded system. Having a mechanism that can enforce program flow can be useful in mitigating attacks such as return oriented programming and stack based code injection. AUTOSAR offers a mechanism that can be re-purposed for CFI, namely in the Watchdog Manager (WdgM). Using the WdgM it is possible to define supervised entities (SE) which are code elements that can be monitored for the order of execution. This results in creating an internal graph of code segments that shall be executed in a specific sequence with time constraints between checkpoints. In Figure 6, we show an example where checkpoints are inserted to allow the WdgM to enforce the program flow of a password checker. If an attacker manages to jump to CP1-2 to unlock the security state, the WdgM will detect a program flow violation because CP1-0 and CP1-1 were bypassed. During the execution of WdgM_MainFunction(), the WdgM detects the violation and can trigger a watchdog reset by not refreshing the watchdog timer. If the attacker chooses to reroute control to his own routine and disables the calling of the WdgM_MainFunction(), the watchdog timer will also trigger a reset because it has not been serviced. Note that one can argue that an attacker can reload the watchdog timer to prevent a reset. This can be made more difficult by only using a windowed watchdog timer. This means the attacker would have to reload the timer in the right time window which is synchronized with the WdgM_MainFunction() call cycle.

7.1.4 OsTiming protections

The locking time protection can be useful in detecting attacks in which a malicious application attempts to disable interrupts for an extended period of time to complete an attack. The inter-arrival time protection can be useful to detect DoS attacks in which an attacker attempts to overwhelm the system by triggering an interrupt too many times to starve the CPU from runtime or to exhaust stack memory resources. An example would be malicious network traffic that results in over triggering the CAN interrupt.

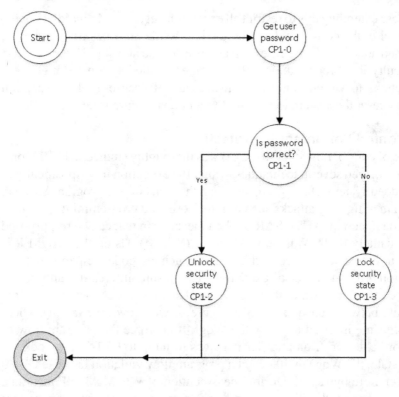

Figure 6   WdgM monitors if password verification checkpoint is skipped to detect an attack.

7.1.5 Hardware resource protection

AUTOSAR OS relies on the MPU to prevent lower ASIL applications from corrupting the data of a higher ASIL application including register settings. One potential use case for security is prohibiting access to the CAN registers by mapping them to a protected MPU entry. This prevents malicious code from directly interacting with the CAN controller to spoof the CAN bus in an infected ECU.

7.2 Security Constraints

As shown in Figure 2, in order to use the safety mechanisms listed above for security, certain constraints are needed to ensure the availability and integrity of those protections.

1. Internal variables of the monitoring software shall be protected against tampering via the MPU partitioning to prevent an attacker from spoofing

their values, for e.g. WdgM global state variable which contains the status of the program flow checks.

2. Software modules executing the security monitoring and enforcement like AUTOSAR OS, WdgM and MCU Driver shall be protected against tampering via secure boot. This is to ensure that those protections are present when the system boots up.

3. Systems that allow runtime integrity checking of software can be leveraged to periodically verify the integrity and authenticity of the software modules executing the protection mechanisms.

4. Configuring OS applications as "Trusted OS Applications" shall be prohibited. The latter has full access to all memory resources which makes it a valuable target for an attacker. Moreover, "Trusted OS Application" is a pure safety characterization which does not imply any security assurance. Thus we recommend that CPU privilege mode is reserved strictly for the OS, while all other tasks and applications shall run in user mode.

5. The software shall be partitioned not only based on safety criticality but also based on security relevance. Having separation between security relevant software and non-security relevant software further enforces the principles of security isolation.

6. In order to raise the difficulty of code injection attacks that can disable protection mechanisms, RAM based execution shall be disabled via the MPU for all RAM partitions (data and stack) in both user and privileged modes. This means that even if an attacker manages to load code on the stack, attempting to execute that code shall result in an MPU exception.

7. Any API call in AUTOSAR that allows disabling a security monitoring shall be disabled, to prevent an attacker from bypassing security checks.

8. The timer interrupt upon which the OS is relying shall be a high priority non-maskable interrupt. This ensures that the OS can execute the security monitoring functions defined in this paper.

9. When using the WdgM for control flow protection, the underlying watchdog timer shall be only configurable once after reset. Attempts to disable the watchdog shall either be ignored or result in a system reset.

7.3 Harmonizing Safety and Security Response

As shown in [29], safe degradation of a safety function can depend on the criticality of the function. Safety critical systems aim to keep alive critical functions for as long as possible when an error is detected. This leads to the definition of different degradation levels that aim to keep alive those functions that are deemed most critical to a vehicle. A function is allowed per the FTTI(fault tolerant time interval) before it has to be degraded due to a fault [27]. Furthermore, AUTOSAR provides flexibility in defining the error response which can be as harmless as a simple notification to the application. From a security point of view, detecting an attack, especially one that has compromised internal memory of the target shall result in a quick response to restore the target to a trusted state. One possible response is to clear the contents of volatile memory and issue a system reset. This can then trigger the secure boot mechanism which checks if the contents of non-volatile memory have been modified. Any such manipulation can result in locking security assets to prevent an attacker from using the affected target to launch attacks on other devices in its direct network. But a sudden reset is not acceptable for safety applications which have explicit requirements against the sudden loss of a safety critical functions like electric power steering. Therefore, there is a need to harmonize the error response to satisfy the safety and security needs of the system. This may result in creating more redundancies in the system to be able to keep alive the safety function even under an attack.

8 Attacks Evaluation

In order to evaluate the attacks described in Section 6, we build a test environment using an automotive grade 32-bit micro-controller, running at a 120 Mhz CPU clock as the test target, and Vector CANalyzer as a simulated attacker. The two are connected together through a CAN FD link with an arbitration baudrate of 500 Kbps and a data rate of 2 Mbps. The identity of the micro-controller is not disclosed in this paper.

8.1 Zero-ID Flood Attack

To simulate the attack, CANalyzer is used to send messages on two CAN FD channels connected together into a single CAN FD channel on the target board. The micro-controller target board controls an RC car by translating the CAN messages into PWM signals that control the steering and driving as shown in Figure 7. This is representative of a malicious attacker that has direct access to the CAN bus where the target ECU resides. The aim is to observe the impact of the zero-ID flood attack on the ability of the target board to steer or drive the car.

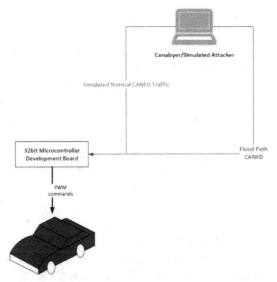

Figure 7 Data Architecture for zero-ID attack.

SOF	11 bit CAN Identifier	r1	IDE	EDL	r0	BRS	ESI	4bit DLC	0-64 bytes: Data Field	21 bit CRC	1	1	1	7 bit EOF	3 INT	IDLE

Figure 8 CAN FD Frame layout, for 64 byte frames CRC is 21 bits long [22].

On the flood path channel, a Communication Application Programming Language (CAPL) script is used to send the zero-ID message in a back to back fashion. On the Normal channel, a CAPL script simulates the drive and steer messages which cycle through a sequence of steering and throttle messages. By enabling the flood attack, control of the RC car becomes very difficult as only a small fraction of control messages is transmitted on the bus. In a real vehicle with timeout monitoring protection, the receiver would simply disable the steering and driving control functions in response to losing messages which would correspond to a successful DoS attack.

8.2 Resource Exhaustion Attack

We first introduce CAN FD which is a relatively new communication protocol that extends the CAN 2.0 standard with a larger payload (up to 64 bytes) and a higher data rate (up to 8 Mbps) [27]. The protocol defines an arbitration baud rate, and a flexible data rate that can be higher than the arbitration rate. This allows CAN 2.0 frames to coexist with CAN FD frames.

In order to evaluate this attack we implemented a reduced AUTOSAR stack that performs the entire chain of CAN message reception and authentication and applied it to a CAN FD network. We assumed that the AUTOSAR CSM [3], is configured in synchronous mode, as a result, SecOC_MainFunction() waits until a buffered secure PDU is authenticated before triggering the next one as shown in Figure 3. The process is repeated until all the buffered secure PDUs have been verified. The attacker was simulated by a software task that runs every 10 ms and produces a variable number of CAN FD messages within a single burst on CAN channel 2 of the micro-controller. A CAPL script in CANalyzer relays the messages from CAN channel 2 to CAN channel 1 to trigger the authentication in the SecOC_MainFunction().

We configured the receiver to process the CAN messages on CAN channel 1 in a 10 ms cycle, i.e. $T_{secoc} = 10$ ms, and configured the CAN controller to receive a maximum of 40 unique messages on CAN channel 1. We also set the SecOCFreshnessCounterSync Attempts value to 1, because the freshness counter is sent in its entirety within the CAN FD frame. This is also meant to increase the attack difficulty, because a larger SecOCFreshnessCounterSync Attempts increases the $T_{processing}$ time needed to verify all the failed MAC values received making the attack easier to succeed. Due to its prevalence in embedded systems, we choose the AES-128 CMAC as the authentication algorithm. Thus the CAN FD frame was constructed to contain 48 bytes of payload data, 8 bytes freshness counter and 8 bytes truncated CMAC. As for the stuff bit factor f, we chose a factor of 15% which is below the maximum value and more biased towards the worst case condition. We then toggle a port pin around the function SecOC_MainFunction() to measure $T_{processing}$ with an oscilloscope.

Using Equations (1) and (2), we can determine that for the parameters of our experiment outlined in Table 2, the maximum burst of messages possible to attack the system is 27 messages with a payload of 64 bytes each. By choosing

Table 2   CAN FD frame time in μs based on 64 byte DLC

Arbitration Rate	500 kbps
Data Rate	2000 kbps
Data Length	64 bytes
Arbitration Time	39.1 μs
Data Rate	314.4 μs
Total Frame time	359.5 μs
SecOC Cycle Time	10 ms
Burst Size	27.81 frames

a 64 byte CAN FD frame length we aim to increase the attack difficulty by minimizing the maximum burst size possible within our time constraint of 10 ms. The next step then is to find the burst size for which $T_{processing}$ exceeds T_{budget}.

Arriving at the execution budget is highly dependent on the application and how the operating system is configured. Typically, the system designer chooses the execution budget of individual tasks based on a static analysis aided by tools that can estimate worst case execution time. For our evaluation since we do not have a real application we assume that SecOC_MainFunction will be among several cyclic functions that are part of the 10 ms Task. Thus we choose the T_{budget} to be 10% of the task cycle time which corresponds to 1 ms. In a real system, execution times can be better estimated based on the demands of the target application. The results of our experiment in Figure 9, show that for $T_{processing}$ to exceed our chosen execution budget of 1 ms, it is

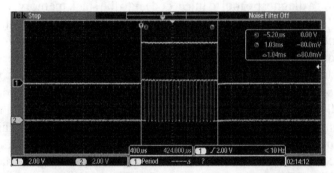

Figure 9   Total runtime is 1.04 ms for processing 22 CAN FD messages of 64 bytes each.

```
void  ApplDiagWriteDataByIdentifier(uint8_t canLen)
{
        uint8_t diagBuffer[8];

        /* this routine copies data from CAN to the
        diagnostic buffer the length in the CAN
        diagnostic request is not being checked this
        allows the buffer to be overrun and the return
        address to be overwritten */

        memcpy(diagBuffer, canBuffer, canLen);
}
```

Figure 10   Buffer overflow routine.

sufficient to send a burst of 22 secure PDUs within a 10 ms cycle. Therefore the number of frames needed to trigger the AUTOSAR OS failure is below the *maxB* = 27 calculated in Table 2 which satisfies Theorem 1. Furthermore, 22 CAN messages is well within the normal number of CAN frames that a typical ECU consumes.

9 Defense Evaluation

For our evaluation of protection mechanisms, we chose the stack usage monitoring protection and performed the evaluation on a RH850 P1x based microcontroller [1] which is ASIL-D qualified.

9.1 Stack Overflow Protection

To demonstrate the security mechanism we create a CAN diagnostic vulnerability that allows a diagnostic tool to inject code into the stack memory reserved for a "trusted" OS application. The diagnostic routine that is loading the data over CAN contains a bug that allows the diagnostic tool to send a well crafted diagnostic request to overflow the stack. The received payload overwrites the return address of the calling routine with a RAM based address which corresponds to the entry point of the malicious routine. Since we do not have access to a full AUTOSAR software stack, we implemented a minimal RTOS that is responsible for setting up the MPU to protect the stack memory against writes by unauthorized software partitions. We also disabled execution rights for stack based addresses to prevent an attacker from fetching instructions from the stack in the case of a successful code injection attack.

   Figure 11 shows the stack state before the attack with the target address to overwrite being 0x4e80. The code listing shown in 10 is a simple routine that contains the buffer overflow exploit we used. Figure 12 shows the contents of the stack after executing the routine which causes the overwrite of the return address with 0xfebff000. Once the routine attempts to return, the CPU pulls the link pointer register value (lp) from the stack causing it to jump to the malicious code shown in Figure 14. The latter attempts to jump to the bootloader after setting the programming flag. The attacker's goal is to bypass

```
0xfebe0f2c   00000000  febe0f84
0xfebe0f34   febe0004  febe0000
0xfebe0f3c   febe0f84  febe0000
0xfebe0f44   00004e80  febe07dc
```

Figure 11   Stack values before executing the attack, the return address is 0x4e80.

```
)xfebe0f34   00000000  55aa55aa
)xfebe0f3c   55aa55aa  55aa55aa
)xfebe0f44   febff000  febe07dc
)xfebe0f4c   febe0fdc  000400db
```

Figure 12   Stack content after executing the attack, the return address is 0xfebff000.

```
.offset 0x0090
#if (MIP_MDP_ENABLE > 0x00000000)
   .extern _MIP_MDP
   jr _MIP_MDP
0x90   .intvect..C.3A.5CCES+0x90:      07805b84          jr        MIP_MDP (0x5c14)
0x94   .intvect..C.3A.5CCES+0x94:      .byte  00,00,00,00
0x98   .intvect..C.3A.5CCES+0x98:      .byte  00,00,00,00
0x9c   .intvect..C.3A.5CCES+0x9c:      .byte  00,00,00,00
```

Figure 13   MPU exception is triggered due to violation of execution rights in RAM.

```
void maliciousRoutine(void)
{
         reprogrammingFlag = C_TRUE;
0xfebff000  maliciousRoutine:            0a01          mov       1, r1
0xfebff002  maliciousRoutine+0x2:        0f4480df      st.b      r1, -32545[gp]
         asm("jr 0x1000"); // jump to the bootloader
0xfebff006  maliciousRoutine+0x6:        07801000      jr        0xfec00006
}
0xfebff00a  maliciousRoutine+0xa:        007f          jmp       [lp]
```

Figure 14   Malicious routine is successfully entered after the stack overflow.

the normal security checks that are required to enter flash programming mode. Due to the security mechanism being active, the vulnerable diagnostic routine is only able to overflow its own stack, but not the stack of other OS applications. Although the stack overwrite is possible, the attempt to fetch the code to jump to the bootloader, immediately triggers an exception as shown in Figure 13.

Note the MPU exception is non-maskable which is important to prevent an attacker from having the ability to delay or block the exception. Once the exception is triggered, the system can log the address where the violation occurred and store that in non-volatile memory for intrusion analysis. The next action is to reset the system which restores the CPU back to its original state. In order to clear the malicious code, it is highly recommended that the CPU always resets the contents of the RAM and stack during startup. Note the above action has to take into consideration the vehicle state to avoid creating a sudden loss of a safety function which would violate a safety goal.

10 Conclusion

In this paper, we presented methods for increasing the security resilience of AUTOSAR based systems. First by considering safety mechanisms as potential attack vectors, we offered several countermeasures that can prevent the abuse of safety mechanisms by attackers. In cases where full prevention was not possible, we proposed an enhanced fail-safe response that factors in malicious attacks as the potential source of the safety failure. Then we analyzed AUTOSAR safety mechanisms that exhibit security properties. We showed that AUTOSAR offered several strong security features if they are used with the proper constraints. We also showed the need to harmonize the response to fault detection to satisfy both the safety and security objectives of the system. In future work we plan on performing our evaluation on a commercially available AUTOSAR stack to produce a set of security requirements that can be integrated with AUTOSAR. We also plan to study the impact of physical attacks on the hardware protections such as RAM and Flash Test features.

References

[1] RH850 P1X Microcontroller Information microcontroller description. Available at: https://tinyurl.com/ybqbbanb [Accessed: 2017-11-28].

[2] Specification of Core Test. AUTOSAR Release 4.2.2

[3] Specification of Crypto Service Manager. AUTOSAR Release 4.2.2

[4] Specification of Diagnostic Event Manager. AUTOSAR Release 4.2.2

[5] Specification of Flash Test. AUTOSAR Release 4.2.2

[6] Specification of Module Secure Onboard Communication. AUTOSAR Release 4.2.2

[7] Specification of Operating System. AUTOSAR Release 4.2.2

[8] Specification of RAM Test. AUTOSAR Release 4.2.2

[9] Specification of Watchdog Manager. AUTOSAR Release 4.2.2

[10] Abadi, M., Budiu, M., Erlingsson, Ú., and Ligatti, J. (2009). Control-flow integrity principles, implementations, and applications. *ACM Transactions on Information and System Security (TISSEC)*, 13, 4.

[11] Bai, Y. (2015). *Practical Microcontroller Engineering with ARM Technology.* John Wiley & Sons.

[12] Checkoway, S., McCoy, D., Kantor, B., Anderson, D., Shacham, H., Savage, S., Koscher, K., Czeskis, A., Roesner, F., Kohno, T., et al. (2011). Comprehensive experimental analyses of automotive attack surfaces. In: *USENIX Security Symposium*, San Francisco.

[13] Cho, K.T., and Shin, K.G. (2016). "Error handling of in-vehicle networks makes them vulnerable," in *Proceedings of the 2016 ACM SIGSAC Conference on Computer and Communications Security*, ACM, 1044–1055.

[14] Day, D.J., and Zhao, Z.X. (2011). Protecting against address space layout randomisation (ASLR) compromises and return-to-libc attacks using network intrusion detection systems. *International Journal of Automation and Computing* 8, 472–483.

[15] Dwoskin, J.S., Gomathisankaran, M., Chen, Y.Y., and Lee, R.B. (2010). "A framework for testing hardware-software security architectures," in *Proceedings of the 26th Annual Computer Security Applications Conference*, ACM, 387–397.

[16] Foster, J.C., Osipov, V., Bhalla, N., and Heinen, N. (2005). *Buffer Overflow Attacks: Detect, Exploit, Prevent.* Syngress Publishing (2005).

[17] Francillon, A., and Castelluccia, C. (2008). "Code injection attacks on harvard-architecture devices," in *Proceedings of The 15th ACM Conference on Computer and Communications Security*, ACM, 15–26.

[18] Francillon, A., Perito, D., and Castelluccia, C. (2009). "Defending embedded systems against control flow attacks," in *Proceedings of The First ACM Workshop on Secure Execution of Untrusted Code*, ACM, 19–26.

[19] Fürst, S., and Spokesperson, A. (2015). AUTOSAR the next generation – the adaptive platform. CARS@EDCC2015.

[20] GbR, A.: Specification of sw-c end-to-end communication protection library.

[21] Glas, B., Gebauer, C., Hänger, J., Heyl, A., Klarmann, J., Kriso, S., Vembar, P., and Wörz, P. (2014). Automotive safety and security integration challenges. In: *Automotive-Safety & Security*, 13–28.

[22] Hartwich, F. (2012). Can with flexible data-rate. *Proc. iCC. Citeseer* (2012).

[23] Lima, A., Rocha, F., Völp, M., and Esteves-Verissimo, P. (2016). "Towards safe and secure autonomous and cooperative vehicle ecosystems," in *Proceedings of the 2nd ACM Workshop on Cyber-Physical Systems Security and Privacy*, ACM, 59–70.

[24] Miller, C., and Valasek, C. (2013). Adventures in automotive networks and control units. *DEF. CON.* 21, 260–264.

[25] Nasser, A.M., Ma, D., and Lauzon, S. (2017). "Exploiting AUTOSAR safety mechanisms to launch security attacks," in *International Conference on Network and System Security*, Springer, 73–86.

[26] Standard, I.: Iso 26262, 2011. Road vehicles Functional Safety (2011).

[27] Standard, I.: Iso 11898, 2015. Road vehicles – Controller area network (CAN) – Part 1: Data link layer and physical signaling (2015).

[28] Tencent: New car hacking research: 2017, remote attack tesla motors again. Keen Security Lab Blog. Available at: https://tinyurl.com/yalxvnoz
[29] Trapp, M., Adler, R., Förster, M., and Junger, J. (2007). Runtime adaptation in safety-critical automotive systems. *Software Engineering*, 1–8.
[30] Wiersma, N., and Pareja, R. (2017). A security assessment of the resilience against fault injection attacks in ASIL-D certified microcontrollers, esCar 2017.

Biographies

Ahmad MK Nasser is a Ph.D. candidate at the University of Michigan Dearborn. He attended Wayne State University where he received his B.Sc. in Electrical Engineering and M.Sc. in Computer Engineering. Ahmad has held various Software Engineering roles throughout his career since 2002 with a focus on basic embedded software and embedded vehicle security. He is a domain expert in embedded systems, flash programming, vehicle diagnostics, communication protocols, AUTOSAR basic software, and hardware based security. He currently works as a senior software manager at Renesas Electronics America, where he leads the secure software center of competence. Ahmad is currently completing a doctorate in Computer Science at the University of Michigan Dearborn. His Ph.D. work centers on the interplay of safety and security in Automotive Systems.

Di Ma is an Associate Professor in the Computer and Information Science Department at the University of Michigan-Dearborn. She also serves as the director of the Security and Forensics Research Lab (SAFE). She is broadly interested in the general area of security, privacy, and applied cryptography. Her research spans a wide range of topics, including smartphone and mobile device security, RFID and sensor security, vehicular network and vehicle security, computation over authenticated/encrypted data, fine-grained access control, secure storage systems, and so on. Her research is supported by NSF, NHTSA, AFOSR, Intel, Ford, and Research in Motion. She received the Ph.D. degree from the University of California, Irvine, in 2009. She was with IBM Almaden Research Center in 2008 and the Institute for Infocomm Research, Singapore in 2000–2005. She won the Distinguished Research Award of the College of Engineering and Computer Science in 2017 and the Tan Kah Kee Young Inventor Award in 2004.

Priya Muralidharan has a Bachelors in Physics and Masters in Information Technology from the University of Delhi, India. She also has a Masters in Computer Science from the University at Buffalo, New York. She is currently working as a Senior Application Engineer at Renesas Electronics America, in the area of Functional Safety. She has over 10 years of experience in embedded software and controls development for various automotive applications such as Electric Power Steering Systems, Hybrid and Electric Vehicles. She has also worked extensively on electronic components such as electric and oil pumps as well as vehicle gateways.

Random Number Generators Based on EEG Non-linear and Chaotic Characteristics

Dang Nguyen, Dat Tran, Wanli Ma and Dharmendra Sharma

Faculty of Science and Technology, University of Canberra,
ACT 2601, Australia
E-mail: Dang.van.Nguyen@canberra.edu.au; dat.tran@canberra.edu.au

Received 1 December 2017; Accepted 13 December 2017;
Publication 2 January 2018

Abstract

Current electroencephalogram (EEG)-based methods in security have been mainly used for person authentication and identification purposes only. The non-linear and chaotic characteristics of EEG signal have not been taken into account. In this paper, we propose a new method that explores the use of these EEG characteristics in generating random numbers. EEG signal and its wavebands are transformed into bit sequences that are used as random number sequences or as seeds for pseudo-random number generators. EEG signal has the following advantages: 1) it is noisy, complex, chaotic and non-linear in nature, 2) it is very difficult to mimic because similar mental tasks are person dependent, and 3) it is almost impossible to steal because the brain activity is sensitive to the stress and the mood of the person and an aggressor cannot force the person to reproduce his/her mental pass-phrase. Our experiments were conducted on the four EEG datasets: AEEG, Alcoholism, DEAP and GrazA 2008. The randomness of the generated bit sequences was tested at a high level of significance by comprehensive battery of tests recommended by the National Institute of Standard and Technology (NIST) to verify the quality of random number generators, especially in cryptography application. Our experimental results showed high average success rates for all wavebands and the highest rate is 99.17% for the gamma band.

Journal of Cyber Security, Vol. 6_3, 305–338.
doi: 10.13052/jcsm2245-1439.634

Keywords: Random number generator, EEG, NIST Test Suite, Security, Cryptography.

1 Introduction

Random number generators (RNGs) are algorithms designed to produce sequences of numbers that appear to be random. RNGs play a crucial role in many applications such as cryptography, machine learning, Monte Carlo computation and simulation, industrial testing and labeling, hazard games, lotteries, and gambling.

In applications where provability is essential, randomness sources (if involved) must also be provably random; otherwise, the whole chain of proofs collapses. In cryptography, due to Kerckhoffs principle where all parts of protocols are publicly known except a secret (the key or other information) that only the sender and the recipient know, it is clear that the secret must not be predictable or calculable by an eavesdropper, i.e., it must be random. For example, the well-known BB84 quantum key distribution protocol [7] would be completely insecure if only an eavesdropper could predict (or calculate) either Alices random numbers or Bobs random numbers or both. From analysis of the secret key rate presented therein, it is obvious that any predictability of random numbers by the eavesdropper would leak relevant information to him, thus diminishing the effective key rate. It is intriguing [34] that in the case that the eavesdropper could calculate the numbers exactly, the cryptographic potential of the BB84 protocol would be zero. Indeed one of the recent successful attacks on quantum cryptography exploits the possibility to control local quantum RNGs by exploiting a design flaw of two commercial quantum cryptographic systems and one practical scientific system. This example shows that the local RNGs assumed in BB84 are essential for its security and may not be exempt from the security proof.

Lotteries are yet another serious business where random numbers are essential. Due to the large sum of money involved (estimated six billion USD annually only online and only in the USA [18]), some countries have set explicit requirements for RNGs for use in online gambling and lottery machines and have set certificate issuing authorities. For example, the Lotteries and Gaming Authority (LGA) of Malta has prescribed a list of requirements for RNGs, stipulated in the Remote Gaming Regulations Act. An RNG that does not conform to this act may not be legally used for gambling business. These rules have been put forward in order to ensure fair game by providers and to prevent possibility that gamers manipulate the system by foreseeing outcomes.

RNGs have been an occupation of scientists and inventors for a long time. Whole branches of mathematics have been invented out of a need to understand random numbers and ways to obtain them. At the dawn of the modern computing era, John von Neumann was one of the first to note that deterministic Turing computers are not able to produce true random numbers, as he put it in his well-known statement that Anyone who considers arithmetical methods of producing random digits is, of course, in a state of sin.

RNGs are one of the hottest topics of research in recent years. There have been about 83 patents per year in the last decade, 1418 in total since 1970, and countless scientific articles published regarding true RNGs. Still, a sharp discrepancy between the number of publications and the very modest number of products (only four quantum RNGs and a handful of Zener noise-based mostly phased-out RNGs) that ever made it to the market clearly indicate the art of immaturity.

Historically, there have been two approaches to random number generation which are algorithmic (pseudo-random) approach and physical process (nondeterministic) one.

Pseudo-random number genrators (RNGs) are also known as deterministic methods that use a mathematical algorithm to produce a long sequence of random numbers. PRNGs have requirements on seed to output a number. If the seed is secret and the algorithm is well-designed, the output number will be unpredictable. Advantages of PRNGs are their low cost, ease of implementation, and user-friendliness, especially in a CPU-available environment such as a PC computer. Moreover, a good RNG should work efficiently, which means that it should be able to produce a large amount of random numbers in a short period of time. For applications like stochastic simulation, stream ciphers, the masking of protocols, and online gambling, huge amounts of random numbers are necessary and thus fast RNGs are required.

While most modern PRNGs pass all known statistical tests, there are myths about some PRNGs being much better than the others. The truth is that every PRNG shows its weakness in some particular application. Indeed PRNGs are often found to be the cause of erroneous stochastic simulations and calculations [11, 13]. As for cryptographic purposes, all major families of PRNGs have been cryptanalyzed so far [32], and use of PRNG where an RNG should be used will therefore present a big security risk for the protocol in question.

For example, the pseudo-random number generator (PRNG) used in Windows operating system is the most commonly used PRNG. The pseudo-randomness of the output of this generator is crucial for the security of almost any application running in Windows. Nevertheless, its exact algorithm was never published. In [12] an attacker can learn future outputs in $O(1)$ time and compute past outputs in $O(2^{23})$ time. These attacks can be run within seconds or minutes on a modern PC and enable such an attacker to learn the values of cryptographic keys generated by the generator. The attacks on both forward and backward security reveal all outputs until the time the generator is rekeyed with system entropy. Given the way in which the operating system operates the generator, this means that a single attack reveals 128 KBytes of generator output for every process.

On the other hand, true random numbers or physical non-deterministic random number generators (TRNGs) seem to be of an ever-increasing importance. Today, true random numbers are most critically required in cryptography and its numerous applications to our everyday life such as mobile communications, e-mail access, online payments, cashless payments, ATMs, e-banking, Internet trade, point of sale, prepaid cards, wireless keys, general cyber security, and distributed power grid security.

However, the characteristics of TRNGs are opposite to those of PRNGs. TRNGs are often biased, this means for example that on average their output might contain more ones than zeros and therefore does not correspond to a uniformly distributed random variable. This effect can be balanced by different means, but this post-processing reduces the number of useful bits as well as the efficiency of the generator. Another problem is that some TRNGs are very expensive or need at least an extra hardware device. In addition, these generators are often too slow for the intended applications. Despite the arguments above, TRNGs have their place in the arsenal. They are used to generate the seed or the continuous input for RNGs.

For implementation, RNGs can be implemented in either hardware or software. Random number generation performed by software utilizes a mathematical algorithm that produces a sequence of statistically independent numbers following a uniform distribution. However, this sequence is deterministic given the algorithm and the seed. While it is possible to implement a mathematical algorithm in hardware and call it a "hardware random number generator", these particular RNGs clearly belong to the category of pseudo-random number generators because they require a seed and produce a deterministic sequence of numbers. True random number generation in hardware depends upon the random characteristics of some physical systems;

for example lava lamps, radioactive decay of atomic nuclei, or noise from a resistor or diode. One of the most important properties of such generators is that they do not need any seed to start producing random sequences.

Biometrics is the science and technology of measuring and statistically analyzing biological data. In information technology, biometrics usually refers to techniques for measuring and analyzing human body characteristics such as fingerprints, eye retinas and irises, voice patterns, facial patterns, and hand measurements, especially for authentication and identification. This kind of application needs to study stationary properties uniquely determined by the attributes of each individual.

Moreover, we believe biometrics have another potential application. Inherent to each biometric measurement is a variability, which is the result of different measurement conditions and ways in which the user presents his or her features to the scanner. This variability effectively represents randomness, which, if extracted, could then be used as a seed for pseudo-random number generators, or directly as a random number sequence: [15, 16, 35]. However, they have some limitations because they are not very secret as expected that can affect to the biometrics-based RNGs in random number regeneration. For example, fingerprint can be changed through human's life, face and iris information can be photographed, and handwriting may be mimicked [22]. Voice can be recorded while a user is speaking. In addition, an adversary can be easy to capture these biometrics by forcing the legitimate user such as threatening with a gun.

In this paper we propose to improve and advance the current state of biometric-based random number generators from EEG source which has disadvantages of low sampling rate and low resolution, but has several advantages: 1) it is noisy, complex and chaotic, and non-linear in nature [1, 28, 31], 2) it is very difficult to mimic because similar mental tasks are person dependent, and 3) it is almost impossible to steal because the brain activity is sensitive to the stress and the mood of the person: an aggressor cannot force the person to reproduce his/her mental pass-phrase [20].

The major contributions of this article are as follows. We explore the randomness of EEG data in order to use them (after codification in integer or bit format) as seeds for pseudo-random number generators or, directly, as random number sequences. We choose the EEG data as a new and promising information source, and develop an algorithm for extracting the inherent randomness. We investigate the use of individual frequency wavebands as random number generators because the wavebands may yield more accurate information about constituent neuronal activities underlying the EEG and consequently, certain changes in the EEG that are not evident in the original full-spectrum EEG may be amplified when each sub-band is analyzed separately [2]. We validate the randomness of four EEG datasets which are Australian EEG, Alcoholism, DEAP, GrazA 2008 datasets. The randomness of the generated bit sequences was then verified at a high level of significance by the standard NIST Test Suite (recommended by the National Institute of Standards and Technology).

These results therefore will open potential possibilities of generating true random numbers for biometrics-based systems to be added to the traditional ones based on physical systems. The main advantage of this alternative in eventual implementations would be, apart from privacy, to provide random bits in real time in a simple and very portable way.

2 Related Work

Since the variability effectively represents randomness, however, biometrics present opportunities for use in random number generation mechanisms as a source of randomness, if such variability can be extracted in a meaningful way.

This concept has been first explored by Sczepanski et al. in [35]. Their method is based on the observation that measurements of physical phenomena yield values that fluctuate randomly in their rightmost decimal digits. The values are partitioned into numbered intervals and then used to generate bits based on interval membership, one bit per value. The authors have tested the method on neurophysical brain signals and galvanic skin response, and statistical tests show that the resulting binary sequences have good randomness properties. The generic construction of the method also makes it possible for the method to be used with any biometrics where the results of the measurements can be quantized to decimal values in a meaningful manner. However, the method requires large datasets of values to generate longer bitstrings, and its security is dependent on the security of the measurement process of the biometrics used. They tested their binary sequences on FIPS statistical tests including four tests: monobit test, poker test, run test, and long run test, Mau-rer universal test and the Lempel-Ziv complexity test. Except the Lempel-Ziv test was been withdrawn by the NIST, other tests belong to the NIST STS that are more difficult to pass.

Moreover, a true random number generator based on the data of fingerprint (FPTRNG) is designed in [39], from the fact that middle grayscale pixels of fingerprint image have large random information. The FPTRNG extracts random information from fingerprint image. It is a true random number generator and is able to produce high quality random number. Its efficiency is high. As the fingerprint image always has large information and the random number generated from it is dedicated to using in fingerprint authentication system, it meets the system requirement for random number. In addition, [15] presents and evaluates the idea of using variability in captured samples could be potentially used to obtain truly random bit sequences, which could be used as a seed for PRNG, or as a random number by itself. The proposed method uses fingerprints as the measured biometric and aims to provide a readily available means of generating random numbers on mobile devices equipped with fingerprint readers.

In [16], strong random numbers are generated from using physical, bio-metric data by investigates how to combine biometric feature extraction and random number generation, how to generate the random numbers and how to verify claimed randomness properties. Simulation results are presented. Their idea presents that, independent from the biometric modality, the only requirement of the proposed solution is feature vectors of fixed length and structure. Each element of such a feature vector is analysed for its reliability, only unreliable positions, which cannot be reproduced coherently from one source, are extracted as bits to form the final random bit sequences. Optionally a strong hash-based random extraction can be used. The practicability presents testing vascular patterns against the NIST-recommended test suite for random number generators.

A recent study [10] indicates that the EEG signals can be treated as indirect random number generator that uses a transformation to output a random number. However, this method only works for input data as integers to generate three-sequence output. Another proposal proposes the use of EEG directly as random number generator in [26]. The limitation of this method may be that it is only applied for the positive real numbers, and this method generates only seven sequences of bits that are quite small.

3 EEG Characteristics

EEG is measurement of the electrical field over the scalp. This electrical field is generated by the synaptic currents within the dendrites of many neurons in the cerebral cortex. The membrane transport proteins pump ions across their neuron membranes and make the neurons electrically charged. This exchanging of ions helps the neuron maintain resting potential and to propagate action potentials. Ions of similar charge repel each other. These ions, after being pushed out, can push their neighbouring ions who in turn push their neighbours. This process creates a wave of ions and is known as volume conduction. When this wave reaches the electrodes on the scalp it creates the difference in voltages between any two electrodes, and can be measured by a voltmeter which gives us the EEG [31].

Because the human head is composed of a number of layers including the scalp, skull, brain and others in between, only the synchronous activity of a large number of active neurons can generate enough potential to be picked up by EEG. EEG activity, therefore, represents a sum of the activity of millions of neurons having similar spatial orientation. Pyramidal neurons of the cortex are well-aligned together. Therefore they are thought to be the main-source EEG signal. By comparison, deep brain activity is more difficult to detect because the voltage fields drop by a factor of the square of distance [31].

3.1 Rhythms of EEG

The amplitudes and frequencies of EEG signals change according to the state of a human such as consciousness or unconsciousness [31]. There are five major brainwave patterns differentiated by their frequency ranges including delta (δ), theta (θ), alpha (α), beta (β), and gamma (γ) (see Figure 1).

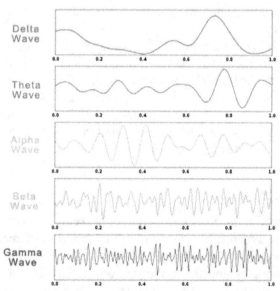

Figure 1   Wavebands example. It can be seen that the gamma and beta wavebands change less widely than others.

- Delta waves: Delta wave frequency lies between 0.5 and 3 Hz, with variable amplitude [24]. Delta waves are associated with deep sleep, and are thought to indicate physical deficiencies in the brain when in the waking state.
- Theta waves: Theta waves are within the range of 4 to 7 Hz, with an amplitude usually greater than 20 μV. Theta waves are caused by emotional stress, such as frustration or disappointment. Theta waves are associated with creativity and deep contemplation [24].
- Alpha waves: The frequency of alpha waves ranges from 8 to 13 Hz and has a voltage ranging from 30 to 50 μV . Alpha waves are associated with relaxed awareness and inattention; however, they seem to indicate an empty mind rather than a relaxed one.
- Beta waves: For beta waves the rate of change lies between 14 and 30 Hz, and has 5–30 μV amplitude [24]. Beta waves are the brainwaves of alertness, and a wakeful state. They are usually associated with logic, analytical and intellectual thinking, active attention and concrete problem solving.
- Gamma waves: Gamma waves lie within the range > 30 Hz and above. Gamma waves usually have low amplitudes, rare occurrence, and relate to left index finger, right toes, and tongue movement [31].

The first human EEG recording was made by Hans Berger in 1924. An EEG recording is obtained by placing on the scalp electrodes which are attached to a cap, net or headset. Conventional EEG sensors use a conductive gel or paste for the electrodes, but many new systems use dry electrodes. These greatly reduce the preparation time, making EEG more accessible to new users [4]. Most applications and studies use a small number of electrodes around the movement-related regions, with locations and names used from the International 10–20 system. Additional electrodes can be added to the standard set-up when a demanding clinical or research application is required. Some high-density arrays can have 256 electrodes. Figure 2 illustrates a typical set of EEG signals in a normal adult brain activity.

Despite several disadvantages, such as low spatial resolution, poor signal-to-noise ratio and an inability to determine neural activities deeply below the cortex, the EEG has a number of advantages. First, hardware costs are significantly lower than other techniques such as Functional magnetic resonance imaging (fMRI), Positron emission tomography (PET) or Magnetoencephalogram (MEG) [37]. Second, EEG sensors are more portable than

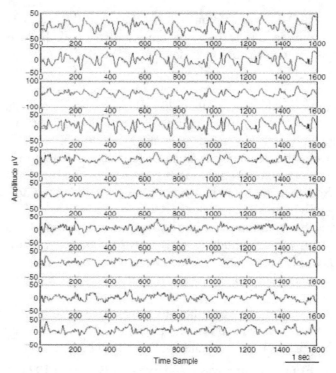

Figure 2   A typical set of EEG signal in a normal adult brain activity [31].

those used in other techniques (fMRI, PET, MEG). Third, the EEG has very-high temporal resolution, usually between 250 and 2000 Hz in clinical and research settings. Finally, the EEG is relatively tolerant of subject movement and methods exist for eliminating movement artifacts in EEG data [25].

3.2 EEG Analysis of Non-linear and Chaotic Characteristics

Linear analysis of EEG signals includes frequency analysis (e.g. Fourier and Wavelet Tranforms) and parametric modeling (e.g. autoregressive models). In general, linear methods can be successfully applied in the study of several problems [3, 5, 14]. However, despite good results have been obtained with linear techniques, they only provide a limited amount of information about the electrical activity of the brain because they ignore the underlying non-linear EEG dynamics. As it is widely accepted, the underlying subsystems of the nervous system that generates the EEG signals are considered non-linear or with non-linear counterparts [38]. Even in healthy subjects, the EEG signals show the chaotic behavior of the nervous system. A brain is also considered a chaotic dynamical system and hence their generated EEG signals are generally chaotic [31, 36].

Besides that, an EEG signal is chaotic in another sense, because its amplitude changes randomly with respect to time. Therefore, due to this non-linear nature of EEG, additional information provided by techniques non-linear dynamics and chaos theory have been progressively incorporated in neurophysiology with the aim to understand the complex brain activity from EEG signals that cannot be measured from linear methods [33]. In particular, non-linear dynamics methods have been used to analyse epilepsy, depth of anesthesia, autism, depression and Alzheimers disease, mental fatigue, brain computer interfaces and emotion recognition [29].

4 NIST Statistical Test Suite

The NIST STS battery consists of 15 empirical tests specially designed to analyse binary sequences (bitstreams). The tests examine randomness of data according to various statistics of bits or statistics of blocks of bits. All NIST STS tests examine randomness for the whole bitstream. Several tests are also able to detect local non-randomness and these tests divide the bitstream into several typically large parts and compute a characteristic of bits for each part. All these partial characteristics are then used for the computation of the test statistic. Each NIST STS test is defined by the test statistic of one of the following three types and examines randomness of the sequence according to:

- bits – these tests analyse various characteristics of bits like proportion of bits, frequency of bit change (runs) and cumulative sums,
- m-bit blocks – these tests analyze distribution of m-bit blocks (m is typically smaller than 30 bits) within the sequence or its parts,
- M-bit parts – these tests analyse complex property of M-bit (M is typically larger than 1000 bits) parts of the sequence like rank of the sequence viewed as a matrix, spectrum of the sequence or linear complexity of the bitstream.

All tests are parametrized by n which denotes the bitlength of a binary sequence to be tested. Several tests are also parametrized by the second parameter denoted by m or M. Since the reference distributions of NIST STS test statistics are approximated by asymptotic distributions (χ^2 or normal), the tests give accurate results (*p*-values) only for certain values of their parameters.

Table 1 The recommended size n of the bitstream for each particular test. Some tests are parameterized by a second parameter m, M, respectively. The table shows meaningful settings for the second parameter and the number of sub-tests executed by each particular test

Test #	Test Name	n	m or M	# sub-tests
1	Frequency	$n \geq 100$		1
2	Frequency within a Block	$n \geq 100$	$M = 128$	1
3	Runs	$n \geq 100$		1
4	Longest run of ones	$n \geq 128$		1
5	Rank	$n \geq 38912$		1
6	Spectral	$n \geq 1000$		1
7	Non-overlapping TM	$n \geq 8m - 8$	$m = 9$	148
8	Overlapping TM	$n \geq 10^6$	$m = 9$	1
9	Maurer's Universal	$n \geq 387840$		1
10	Linear Complexity	$n \geq 10^6$	$M = 500$	1
11	Serial		$m = 16$	2
12	Approximate Entropy		$m = 10^6$	1
13	Cumulative	$n \geq 100$		2
14	Random Excursions	$n \geq 10^6$		8
15	Random Excursions Variant	$n \geq 10^6$		18

Table 11 summarizes appropriate values of the parameters for each particular test recommended by NIST [30].

Several of the NIST STS tests are performed in more variants, i.e., they execute several sub-tests and examine more properties of the sequence of the same type. For instance, the Cumulative sum test examines a sequence according to forward and backward cumulative sum. Table 1 also summarizes the number of sub-tests performed by each particular test. The Non-overlapping template matching test is marked by an asterisk since the number of its sub-tests is not fixed and depends on the value chosen for the parameter m (the number 148 mentioned in the Table 1 corresponds to the default value of the parameter $m = 9$).

There are several ways to interpret a set of p-values computed by an empirical test of randomness. NIST adopted the following two ways:

- The examination of the proportion of sequences that pass a certain statistical test relative number of sequences passing the test should lie within a certain interval.
- The uniformity testing of p-values: p-values computed for random sequences should be uniformly distributed on the interval $[0; 1)$. Uniformity of p-values can be tested again using statistical tests (uniformity of p-values forms a hypothesis).

4.1 Proportion of Sequences Passing Tests

The probability that a random sequence passes a given test is $1 - \alpha$ that is equal to the complement of the significance level α. For multiple random sequences, the ratio of passed sequences to all sequences in a given test is usually different but close to $1 - \alpha$ and hence should fall into a certain interval around $1 - \alpha$ with a high probability. The interval is computed using the significance level α as

$$1 - \alpha \pm \sqrt{\frac{\alpha(1 - \alpha)}{k}} \tag{1}$$

where k is the number of tested sequences. The acceptable ratio of passed sequences to all tested sequences should fall within the interval 0.99 ± 0.02985 for the significance level $\alpha = 0.01$ and the total number of sequences $k = 100$. It means that if a data set of 100 sequences is tested in a NIST test such as Frequency test that results in more than 4 failed sequences, this data set is marked by an asterisk (*). This data set is considered as non-random for the examination of the proportion of passing tests if it has more than 7 asterisks (4%) [21].

We define the success rate in a test as the ratio of passed sequences to all tested sequences, and the failure rate as the ratio of failed sequences to all tested sequences in the NIST tests.

4.2 Uniformity of *p*-values

The *p*-values computed by a singe test should be uniformly distributed on the interval [0,1), and can be interpreted simply as: "the probability that a perfect random number generator would have produced a sequence less random than the sequence that was tested" [30]. For each statistical test, a set of *p*-values (corresponding to the set of sequences) is produced. The NIST uses one sample χ^2 test to assess the uniformity of *p*-values. χ^2 test measures whether the observed discrete distribution (histogram) of some feature follows the expected distribution. The χ^2 test works well only for $k/10$ greater than 5.5. Therefore, the number of tested sequences should be at least 55 ($k \geq 55$) to get a meaningful result for the uniformity test. For a fixed significance level, a certain percentage of *p*-values are expected to indicate failure. The NIST STS documentation recommends a very small value for the significance level $\alpha = 0.0001$ for the uniformity test, i.e., *p*-values are considered as non-uniform if a *p*-value is smaller than 0.0001, and being marked by an asterisk (*). An α of 0.0001 indicates that one would expect one sequence in 10000 sequences to be rejected by the test if the sequence was random. For a p-value ≥ 0.0001, a sequence would be considered to be random with a confidence of 99.99%. For a p-value < 0.0001, a sequence would be considered to be non-random with the same confidence. The dataset of 100 sequences is considered as non-random for the uniformity testing if it has more than 3 asterisks (1.60%) for $\alpha = 0.0001$ [21].

We define a non-uniformity rate as the ratio of non-uniform sequences to all tested sequences in the NIST STS. We aim to have a trade-off between the average of success rate, the failure rate and the non-uniformity rate on all tests.

5 Proposed Method of Binary Sequence Generation

According to the study conducted in [10], the EEG signal cannot be used directly as a source of random number generator and a transformation is required to transform the EEG signal into random sequences of bits. We verified the changing pattern frequency of four EEG datasets (described in Section 6) in the same way in [10]. As shown in Figure 3, the results are similar

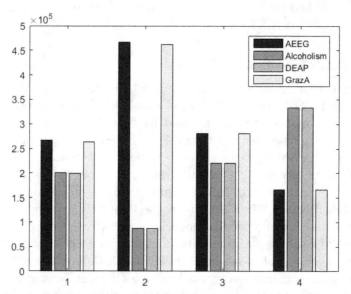

Figure 3 The frequency of down–down, down–up, up–down, and up–up in the EEG signals of all four datasets. The labels 1, 2, 3, 4 on the x-axis correspond to four datasets: AEEG, Alcholism, DEAP, GrazA 2008 respectively. Apparently, these four patterns are not uniformly distributed, so EEG signals cannot be good RNGs.

to [10] in which the frequency of four changing patterns, including up-up up-down, down-up, and down-down, are not uniformly distributed. Therefore, the EEG signal cannot be used directly as random number generators. This is clear proof that the EEG signal requires the transformation before used as random number generators.

We have conducted a similar approach to [10] on the Alcoholism dataset [6]. However, we found more failed tests than those presented in [10]. This could be the fact that our EEG dataset is quite different from the datasets in [10] that contain both negative and positive real numbers. Our EEG signals are measured in floating point values and also have small magnitude values. To handle the EEG signals in our case, we propose a new method in order to improve the success rate in randomness testing as follows. Let vector x be an original EEG signal sequence of n real number sample values:

$$x = (x_1 \ldots, x_n) \text{ with } x_i \in \mathbb{R}, i = 1, \ldots, n \qquad (2)$$

In order to access the fluctuations in the EEG data, we multiply the original EEG data by 10^d to keep significant precision up to d precision floating-point value to obtain integer value of EEG data. The value of d is based on the number of digits in fractional part from EEG raw data. Finally, we perform bit shift operation of b to the right. All these operations can be expressed as follows:

$$y_i = (x_i \times 10^d) \gg b \in \mathbb{Z}, \text{ with } i = 1, \ldots, n \qquad (3)$$

Finally, we compute binary sequences z as:

$$z_i = \lfloor y_i \mod 2 \rfloor \text{ with } i = 1, \ldots, n \qquad (4)$$

It is noted that y_i is a real number after performing the bit shift operation in Equation (3), thus the modular operation $y_i \mod 2$ results in a real number.

6 EEG Datasets

6.1 Australian EEG Dataset (AEEG)

The Australian EEG Database is a collaboration project between the University of Newcastle and the John Hunter Hospital to convert 18,500 hospital EEG records into a web-based searchable database that takes 2 years to complete. The database consists of EEG data of patients, ranging from premature infants to people aged over 90 years.

The EEG data set was recorded using 3 common montages (neonate, infant and adult), bipolar connections and standard International System 10–20 electrode placements. Recordings were undertaken in the resting state with eyes open and eyes closed [17].

A subset of this Australian EEG dataset was used in this paper. It consists of EEG recordings of 80 patients. The recordings were downloaded with the search criteria that the recordings come from both men and women of various ages. The recordings were made by 23 electrodes placed on the scalp, sampled at 167 Hz for about 20 minutes.

6.2 Alcoholism

The Alcoholism datasets come from a study of EEG correlates of genetic predisposition to alcoholism Begleiter [6]. The datasets contain EEG recordings of control and alcoholic subjects. Each subject was exposed to either a single stimulus (S1) or to two stimuli (S1 and S2); which were pictures of objects chosen from the 1980 Snodgrass and Vanderwart picture set. When two stimuli were shown, they were presented in either a matched condition (where S1 was identical to S2) or in a non-matched condition (where S1 differed from S2). The 64 electrodes placed on the scalp were sampled at 256 Hz for 1 second.

There are three versions of the EEG datasets at different sizes: small, large and full datasets. This study uses the full dataset that contains 122 subjects with 120 trials for each.

6.3 BCI-Competition Graz Datasets

The Graz dataset A (GrazA 2008) in the BCI Competition 2008 comes from the Department of Medical Informatics, Institute of Biomedical Engineering, Graz University of Technology for motor imagery classification problem in BCI Competition 2008 [9].

The GrazA 2008 dataset consists of EEG recordings from 9 subjects. The subjects were right-handed, had normal or corrected-to-normal vision, and were paid to participate in the experiments. All volunteers sat in an armchair, and watched a screen monitor placed approximately 1 metre away at eye level. The recording was made with a 64-channel EEG amplifier from Neu-roscan at 250 Hz with time length 7 seconds for each trial. The GrazA dataset consists of two sessions on different days with, 288 trials per session. Each subject was required to do 4 motor-imagery tasks (left hand, right hand, foot, tongue).

Twenty-two Ag/AgCl electrodes were used and the signals were bandpass-filtered between 0.5 Hz and 100 Hz. Each subject carried out left or right hand-motor imagery on two different days within two weeks. Each session consisted of six runs with ten trials each and two classes of imagery.

6.4 DEAP Dataset

DEAP is a dataset for emotion analysis using EEG (Dataset for Emotion Analysis using Electroencephalogram, Physiological and Video Signals) which is an open database proposed by Koelstra et al. [19]. EEG signals of 32 participants were recorded while they watched 40 one-minute long excerpts of music videos. Participants rated each video in terms of the levels of arousal, valence, like/dislike, dominance, and familiarity. The dataset includes two parts:

- The ratings from an online self-assessment, where 120 one-minute extracts of music videos were each rated by 14–16 volunteers based on arousal, valence and dominance.
- The participant ratings, physiological recordings and face video of an experiment in which 32 volunteers watched a subset of 40 of the above music videos. EEG and physiological signals were recorded and each participant also rated the videos as above.

The data set was recorded at a sampling rate of 512 Hz in two separate locations with 40 channels. The first 22 participants were recorded in Twente and the remaining in Geneva. Then the data was downsampled to 128 Hz, and segmented into 60 second trials and a 3 second pre-trial baseline removed.

6.5 Summary

The summary of those datasets is listed in Table 2, and Figure 4 shows the sample EEG signals of these datasets. In this figure, the four rows of signals from top to bottom are samples EEG signals from the AEEG, Alcoholism, DEAP and GrazA 2008 datasets, respectively.

Table 2 A brief description of EEG datasets

Datasets	#Subjects	#Trials	#Sessions	Length (s)	Sampling (Hz)	#Channel
Australian EEG	80	1	1	1200	167	23
Alcoholism	122	120	1	1	256	64
DEAP	32	40	1	60	128	40
GrazA 2008	9	288	2	7.5	250	22

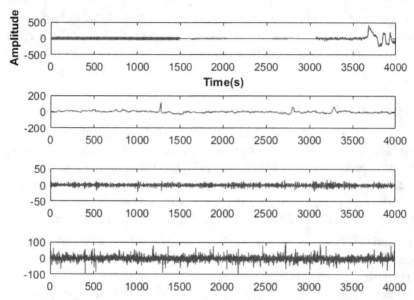

Figure 4   The sample EEG signals of four datasets. From top to bottom are examples from the AEEG, Alcoholism, DEAP and GrazA 2008 datasets, respectively. It can be seen that the EEG signal in GrazA 2008 dataset changes more wildly than other EEG signals.

7 Experimental Results

We carefully set up our experiments to be the two-factor experiments for a number of samples and a number of digits. Some parameters of the proposal can greatly affect the performance, setting different values on them jeopardizes the credibility of experiments. Therefore, we ensure the parameters of the method for all EEG datasets to be the same. For a number of digits, the Alcoholism and Australian EEG datasets adopted a constant number of $d = 3$, and the DEAP and GrazA 2008 adopted a constant of $d = 4$ due to the natural originality of their EEG raw data. For another factor, we first removed all subjects of each dataset that did not have enough samples to produce a single binary sequence of a million bits. For the remaining subjects, we combined its number of channels in a one-second EEG sample into a single set, and joined these single sets into an EEG sequence. Then, we used a suitable number of samples to generate a single binary sequence that contained a million bits. For an example of the Alcoholism dataset, we combined 64 channels in an one-second EEG sample into a single set for each subject. There were 112 subjects who had 62 one-second EEG samples, and we joined those 62 single

Table 3 Parameter settings for our experiments

Datasets	Input				Output	
	#Subjects	#Samples	#Channels	d	#Bits	#Sequences
Alcoholism	122	62	64	3	1015808	112
AEEG	80	261	23	3	1002501	78
DEAP	32	196	40	4	1003520	32
GrazA 2008	9	182	22	4	1001000	9

sets into an EEG sequence for each of them. The 10 subjects left did not have enough 62 samples, so finally we produced 112 EEG sequences. Each EEG sequence was long enough to generate a single binary sequence that contained $62 \times 64 \times 256 = 1015808$ bits. Setting parameters for NIST STS was the same as described in Section 4. Table 3 summarizes the experimental set up:

In order to verify the randomness of EEG data, we conducted a number of experiments for all of EEG datasets in accuracy comparison of EEG and its five wavebands including alpha, beta, delta, gamma and theta. We also examined the proportion of passing sequences and tested the uniformity of p-values using three metrics: the success rate, the failure rate and the non-uniformity rate in which the non-uniformity rate was computed for 2 datasets due to a small number of sequences in the GrazA and DEAP datasets. We aimed to balance these metrics to maximize the success rate (as high as possible), and to minimize both the failure rate (<4.00) and non-uniformity rate (<1.60). We conducted the experiments in Matlab R2015b on a DELL PC with i5-5200U 2.20 GHz processors, 8.00 GB memory, and Windows 7.0 operating system.

7.1 Can EEG be Random Number Generators?

We firstly tested our method for the EEG signal to investigate whether EEG can be used directly as random number generator. Since the method is based on the number of right-shift bits, we varied the value of b from 1 to the maximum bit of 11. We used the EEG datasets which has the resolution of 11 bits to experiment the effect of b on randomness through by the success rate, the failure rate and the non-uniformity rate. We expected to find out a number of right-shift bits that balances the three rates. Figure 5 demonstrates a variation of these rates on the four datasets, and it shows that these values differ from EEG datasets: 9 for GrazA, 1 for DEAP, 7 for AEEG and 4 for Alcoholism. While two curves of the success and failure rates are nearly constant of high values for the GrazA (Figure 5(a)), the success rate decreases slowly, but the failure

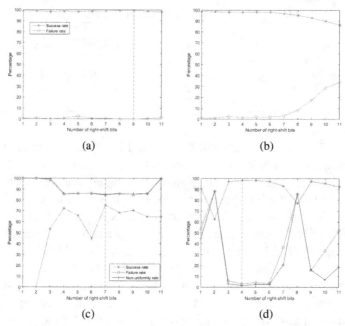

Figure 5 A change of success and failure rates on different right-shift bits: (a) GrazA, (b) DEAP, (c) AEEG and (d) Alcoholism. GrazA and DEAP datasets do not have *p*-values, but AEEG and Alcoholism have as they contain more than 55 tested sequences.

rate goes down quickly when *b* increases for the DEAP (Figure 5(b)). On the other hand, these curves fluctuate for the AEEG, and especially, "randomly" for Alcoholism datasets. The reason can be from the chaotic and complex characteristics of EEG signals [31].

Table 4 summarizes the results of statistical tests for the EEG signal that indicates the randomness of GrazA and DEAP datasets. However, the DEAP dataset failed on two important tests including block frequency and FFT, and

Table 4 Statistical results of EEG at the trade-off of three rates. The Alcoholism and AEEG have high non-uniformity rate, and the DEAP failed on two important tests of Block frequency and FFT

Datasets	*b*	Success Rate	Failure Rate	Non-uniformity Rate
GrazA	9	99.52	0.00	–
DEAP	1	98.69	1.06 (Block frequency, FFT)	–
AEEG	7	75.18	85.57	85.11
Alcoholism	4	98.21	3.19	1.60

hence it was not good random. Therefore, only GrazA 2008 can be used as a RNG. This result is similar to others in [10, 27]. It means that EEG signal will be a RNG if a small number of sequences generated but not on other cases.

In the next experiment, we performed our method for the five wavebands to investigate which waveband can be used directly as RNG. Figure 6 shows their performances. The row at bottom reporting for the theta band contains only three figures because the implementation of this band is dumped due to bad sequences generated when $b = 9$, so we do not present its results here for the Alcoholism dataset.

From Figure 6, it can be seen that except for the AEEG dataset at a number of digits $b > 9$ and DEAP with the delta band, the success rates for the proposed method are higher than 95%, and the failure and non-uniformity rates are lower 5%. The score gaps among different EEG datasets are also very small. This indicates that the proposed method is relatively effective in dealing with the datasets with separated tasks: GrazA 2008 of motor imaginary, DEAP of emotion, AEEG of epilepsy, and Alcoholism of alcohol. Figure 6 also shows that these sub-bands perform a little difference in which the gamma and beta bands seem to be the two best of performance (the second and fourth rows of curves from the top). The reason could be that these two sub-bands are less chaotic and complex than others [31], and our proposed method is deterministic and stable. As a result, the less chaotic and complex characteristics of the gamma and beta bands make lower variation that leads to higher performance.

7.2 Optimization of *b*

In this subsection, we investigate the number of digits b to maximize the success rate and minimize the failure and non-uniformity rates for each of sub-band. Table 5 lists scores of these three rates on different EEG datasets with different wavebands. Based on this table, it is hard to optimize b. For example of the alpha band, there are three different options for b being 8 of GrazA 2008, 3 of DEAP and 1 of AEEG and Alcoholism. Therefore, we performed a computational approach to optimize b as follows. We first considered a case for the success rate and the other two rates are computed the same. For each sub-band, we first grouped all scores of the success rate for 4 datasets into a table. Then, we computed an average of those as seen in Table 6 for an example of the alpha band. We finally sorted them by the order of success rate, failure rate and non-uniformity rate.

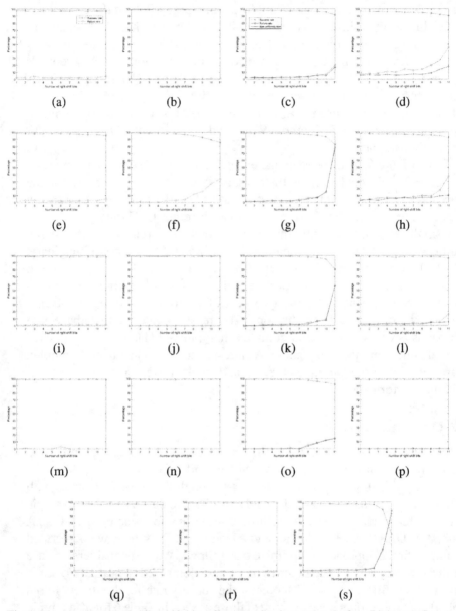

Figure 6 A summary of right-shift bits effects on the three rates. Left to Right: (a) GrazA 2008, (b) DEAP, (c) AEEG and (d) Alcoholism, respectively. Two left figures have only two curves because GrazA 2008 and DEAP datasets does not have *p*-values as they contains less than 55 tested sequences. Top to Bottom: Alpha, Delta, Beta, Gamma and Theta, respectively.

Table 5 Statistical results of the five EEG datasets

		b	1	2	3	4	5	6	7	8	9	10	11
GrazA	Alpha	SR	97.64	97.28	96.57	97.34	97.20	97.16	96.93	97.70	**96.93**	97.10	97.00
		FR	2.13	2.13	4.26	2.13	2.13	2.13	2.13	2.13	**2.66**	2.66	4.26
	Delta	SR	**98.64**	97.93	97.80	97.93	98.05	98.00	97.70	97.34	96.69	96.72	95.33
		FR	**1.60**	3.19	3.19	2.66	2.13	2.13	2.13	2.13	3.19	2.66	5.32
	Beta	SR	**99.17**	99.05	98.88	98.52	98.70	98.62	97.93	98.48	98.35	97.86	98.23
		FR	**1.06**	1.06	1.60	1.60	1.06	1.06	2.13	1.60	1.60	2.13	1.60
	Gamma	SR	99.14	98.98	98.86	99.17	**99.53**	98.68	98.88	98.70	99.26	98.90	98.51
		FR	0.00	1.06	0.00	0.00	**0.00**	3.19	0.00	0.00	0.00	0.00	0.00
	Theta	SR	**97.34**	97.28	96.92	96.40	97.34	96.87	96.98	96.99	96.69	96.51	96.30
		FR	**2.13**	2.66	2.13	2.13	2.13	2.13	3.19	2.13	2.13	4.26	3.72
DEAP	Alpha	SR	98.94	98.88	**98.99**	98.78	98.75	98.76	98.91	98.80	98.64	97.98	98.59
		FR	0.53	0.53	**0.53**	0.53	1.06	1.06	0.53	0.53	0.53	1.06	1.06
	Delta	SR	**98.86**	98.68	98.75	98.50	97.88	97.86	96.90	95.17	92.68	89.30	85.45
		FR	**0.53**	1.06	0.53	0.53	2.66	3.19	4.26	11.70	15.96	27.13	33.51
	Beta	SR	99.12	98.66	99.00	98.92	99.00	**99.18**	98.60	98.66	99.09	99.00	98.72
		FR	0.00	0.00	0.00	0.00	0.00	**0.00**	0.00	0.00	0.00	0.00	0.53
	Gamma	SR	99.22	98.96	98.86	99.02	98.72	98.70	99.08	99.15	99.12	99.07	**99.25**
		FR	0.00	0.00	0.00	0.00	0.00	0.00	0.00	0.00	0.00	0.00	**0.00**
	Theta	SR	98.58	98.50	98.55	98.04	98.55	98.80	98.43	98.16	98.36	98.37	98.32
		FR	0.53	0.53	1.06	1.06	1.60	0.53	0.53	0.53	1.06	1.60	1.60
AEEG	Alpha	SR	**97.86**	97.53	97.62	97.38	97.22	97.17	97.17	97.09	96.55	95.62	91.45
		FR	**2.13**	3.19	2.13	2.13	3.19	2.66	3.72	3.19	6.38	6.91	19.68
		NUR	2.66	**2.13**	2.13	2.13	2.66	2.66	3.19	4.26	5.32	5.32	17.02
	Delta	SR	**98.72**	98.27	98.44	98.19	98.06	97.92	97.60	96.68	95.60	93.51	82.56
		FR	**1.59**	1.59	1.59	3.19	2.13	2.13	4.26	6.38	7.45	15.43	77.13
		NUR	**1.59**	1.59	1.59	2.13	2.66	2.13	2.66	5.32	6.91	14.89	78.72
	Beta	SR	**99.01**	98.84	98.98	98.70	98.88	98.96	98.73	98.08	97.22	94.95	80.27
		FR	**0.00**	0.00	1.06	0.53	0.53	1.59	1.06	4.26	5.85	8.51	56.38
		NUR	**0.00**	0.00	0.53	1.59	1.59	2.13	1.59	2.13	6.38	7.45	56.38
	Gamma	SR	98.90	**99.04**	99.03	98.98	98.93	98.96	98.99	97.92	96.10	94.50	91.89
		FR	0.00	**0.00**	0.53	0.00	0.00	1.06	0.00	5.31	8.51	12.23	14.89
		NUR	0.00	**0.00**	0.00	0.00	0.00	0.00	0.00	3.72	7.98	11.70	14.36
	Theta	SR	**97.67**	97.29	97.36	97.19	96.77	96.99	97.00	96.64	96.48	89.17	54.30
		FR	**2.13**	2.13	2.13	2.66	3.72	3.19	3.19	4.26	5.85	30.85	87.23
		NUR	**2.13**	2.13	2.13	3.19	3.19	3.19	2.66	3.19	5.32	32.45	82.98
Alcoholism	Alpha	SR	**96.89**	96.66	96.29	96.15	95.86	95.32	95.04	94.63	93.96	92.26	90.65
		FR	**7.45**	7.98	8.51	11.17	10.11	14.36	12.77	14.36	19.68	27.13	45.74
		NUR	**5.85**	6.38	5.85	6.91	5.85	6.38	7.45	6.91	9.04	13.83	18.09
	Delta	SR	**98.11**	97.85	97.58	97.45	97.35	97.12	96.76	96.10	95.99	95.08	92.53
		FR	**3.72**	4.79	6.38	7.45	6.38	7.45	8.51	10.11	9.57	17.55	37.23
		NUR	**3.19**	4.26	3.19	5.85	5.85	6.91	6.38	7.45	7.45	9.57	10.64
	Beta	SR	98.52	98.26	**98.62**	98.41	97.32	98.47	98.40	98.26	98.11	97.92	96.25
		FR	3.19	2.13	**2.13**	3.19	3.19	2.13	3.72	3.19	4.79	5.32	15.96
		NUR	**1.60**	1.60	2.13	2.13	2.13	2.13	2.13	2.13	4.26	4.26	5.32
	Gamma	SR	98.80	98.70	98.94	**99.02**	98.94	98.80	98.87	99.01	98.89	98.85	98.78
		FR	0.00	0.00	0.00	**0.00**	0.00	0.00	0.53	0.00	0.00	0.53	0.00
		NUR	0.00	0.00	0.00	**0.00**	0.00	0.00	0.00	0.00	0.00	0.00	0.00
	Theta	–											

SR = success rate, FR = failure rate and NUR = non-uniformity rate.

Table 6   The success rate of alpha band for optimizing the number of digits b. The highest of success rate is at $b = 1$

b	1	2	3	4	5	6	7	8	9	10	11
GrazA	97.64	97.28	96.57	97.34	97.20	97.16	96.93	97.70	96.93	97.10	97.00
DEAP	98.94	98.88	98.99	98.78	98.75	98.76	98.91	98.80	98.64	97.98	98.59
AEEG	97.86	97.53	97.62	97.38	97.22	97.17	97.17	97.09	96.55	95.62	91.45
Alcoholism	96.89	96.66	96.29	96.15	95.86	95.32	95.04	94.63	93.96	92.26	90.65
Average	**97.83**	97.59	97.37	97.41	97.26	97.10	97.01	97.06	96.52	95.74	94.42

Table 7   Optimizing results of the five wavebands in terms of the average of success rate, failure rate and non-uniformity rate (%). The optimization of b is at 1 for alpha, delta, theta, and beta, and at 4 for gamma

	b	1	2	3	4	5	6	7	8	9	10	11
Alpha	SR	**97.83**	97.59	97.37	97.41	97.26	97.10	97.01	97.06	96.52	95.74	94.42
	FR	**3.06**	3.46	3.86	3.99	4.12	5.05	4.79	5.05	7.31	9.44	17.69
	NUR	**4.26**	4.26	3.99	4.52	4.26	4.52	5.32	5.59	7.18	9.57	17.55
Delta	SR	**98.58**	98.18	98.14	98.02	97.84	97.72	97.24	96.32	95.24	93.65	88.97
	FR	**1.86**	2.66	2.92	3.46	3.33	3.72	4.79	7.58	9.04	15.69	38.30
	NUR	**2.39**	2.92	2.39	3.99	4.26	4.52	4.52	6.38	7.18	12.23	44.68
Beta	SR	**98.96**	98.70	98.87	98.64	98.48	98.81	98.42	98.37	98.19	97.43	93.37
	FR	**1.06**	0.80	1.20	1.33	1.20	1.19	1.73	2.26	3.06	3.99	18.62
	NUR	**0.80**	0.80	1.33	1.86	1.86	2.13	1.86	2.13	5.32	5.85	30.85
Gamma	SR	99.02	98.92	98.92	**99.05**	99.03	98.79	98.96	98.70	98.34	97.83	97.11
	FR	0.00	0.27	0.13	**0.00**	0.00	1.06	0.13	1.33	2.13	3.19	3.72
	NUR	0.00	0.00	0.00	**0.00**	0.00	0.00	0.00	1.86	3.99	5.85	7.18
Theta	SR	**97.86**	97.69	97.61	97.21	97.55	97.55	97.47	97.26	97.18	94.68	82.97
	FR	**1.60**	1.77	1.77	1.95	2.48	1.95	2.30	2.31	3.01	12.23	30.85
	NUR	**2.13**	2.13	2.13	3.19	3.19	3.19	2.66	3.19	5.32	32.45	82.98

Table 7 summarizes the results for all the five sub-bands. The table indicates that the optimization value of b is at 1 for the delta, theta, and at 4 for the gamma. For the two remaining cases, we select the optimization at 1 instead of 2 based on the following two reasons. The first reason is from an observation that two out of three rates are the best at 1, and the difference is very small (less than 0.3%). Another reason is from a study on Table 5 that the scores of the alpha and beta at $b = 1$ is better than at $b = 2$: higher on the success rate, and lower on others. The reasons of optimization can be as follows: (1) the frequencies of four wavebands (alpha, beta, delta and theta) are small and the amplitudes are short, (2) the frequency of gamma band is high at more than >30 Hz, and its amplitude is also long in comparison to other wavebands that results in larger values of gamma data. Therefore, the optimizing value of b is higher for gamma band than for others.

7.3 Randomness Testing

Table 8 shows a summary of scores of five wavebands at the optimised values of b. The table indicates that most of these wavebands have high proportion of passing tests, and low proportion of failing uniformity tests, except the alpha for Alcoholism dataset. While the results for gamma and beta bands can be considered random because all of the failure rates are smaller than 4%, and most of non-uniformity rates are less than 1.60%, the alpha, delta and theta bands cannot be considered as good random.

In addition, the results of statistical Test Suite are shown in Tables 9 and 10 for these wavebands. These results show that EEG signal achieves the worse performance, and three wavebands including alpha, delta and theta do not pass frequency, block frequency, approximate entropy, universal and serial tests that are some of the important tests, and it also supports the conclusion that they cannot be considered as random. For the beta band, the average success rate is very high at 98.96%, and the average failure rate is less than 4%. However, there is a case of Alcoholism dataset that the beta is not good random because it failed on three important tests (1.60%) for uniformity testing: block frequency, approximate entropy and serial. Therefore, it needs further investigation on the beta band.

Table 8 Comparison of EEG signal and its five wavebands in terms of the average of three rates (%). The gamma band achieves the highest success rate, and the lowest failure rate and non-uniformity rate

		GrazA	DEAP	AEEG	Alcoholism	Average
SR	EEG	98.82	98.10	72.24	98.21	91.84
	Alpha	97.64	98.94	97.86	96.89	97.83
	Delta	98.64	98.86	98.72	98.11	98.58
	Beta	99.17	99.12	99.01	98.52	98.96
	Gamma	99.17	99.02	98.98	99.02	**99.05**
	Theta	97.34	98.58	97.67	–	97.86
FR	EEG	1.06	1.60	85.11	3.19	22.74
	Alpha	2.13	0.53	2.13	7.45	3.06
	Delta	1.60	0.53	1.60	3.72	1.86
	Beta	1.06	0.00	0.00	3.19	1.06
	Gamma	0.00	0.00	0.00	0.00	**0.00**
	Theta	2.13	0.53	2.13	–	1.60
NUR	EEG	–	–	86.17	1.60	43.88
	Alpha	–	–	2.66	5.85	4.26
	Delta	–	–	1.60	3.19	2.40
	Beta	–	–	0.00	1.60	0.80
	Gamma	–	–	0.00	0.00	**0.00**
	Theta	–	–	2.13	–	2.13

Table 9 Summary of statistical tests for passing test in terms of failure rates. The proportion of failing tests of gamma band is at 0%

Datasets	Statistical Test	Sub-bands				
		Alpha	Beta	Delta	**Gamma**	Theta
GrazA 2008	Block Frequency	X	X	X		X
	Non overlapping Templates					
	Approximate Entropy	X				X
	Serial	X	X	X		X
DEAP	Block Frequency	X		X		X
	FFT					
AEEG	Block Frequency	X		X		X
	Approximate Entropy	X				X
	Random Excursions					
	Serial	X		X		X
	Frequency					
Alcoholism	Block Frequency	X X	X	X		–
	Cumulative sums (Forward)					–
	Cumulative sums (Reverse)					–
	Runs	X				–
	FFT					–
	Non overlapping Templates	X		X		–
	Universal	X		X		–
	Approximate Entropy	X	X	X		–
	Serial	X	X	X		–

Table 10 Summary of statistical tests for uniformity in terms of non-uniformity rates. The uniformity tests of gamma band is at 100% of passing

Datasets	Statistical Test	Sub-bands				
		Alpha	Beta	Delta	**Gamma**	Theta
AEEG	Block Frequency	X		X		X
	Approximate Entropy	X				X
	Random Excursions					
	Serial	X		X		X
	Block Frequency	X		X	X	–
Alcoholism	Cumulative sums (Forward)	X				–
	Cumulative sums (Reverse)					–
	Runs	X				–
	Non overlapping Templates	X		X		–
	Universal	X				–
	Approximate Entropy	X	X	X		–
	Serial	X	X	X		–

Table 11 Comparison on different methods for 112 sequences tested. *p*-values are all zero for three methods

Statistical Test	Proportion of Passing Sequences		
	Proposed Method	Bum Bum Shub	Micali Schnorr
Frequency	108/112	112/112	111/112
Block Frequency	112/112	111/112	112/112
Cumulative Sums (Forward)	109/112	112/112	111/112
Cumulative Sums (Reverse)	108/112	112/112	112/112
Runs	107/112	111/112	111/112
Longest Runs of Ones	112/112	110/112	111/112
Rank	111/112	110/112	109/112
FFT	111/112	109/112	112/112
Non Overlapping Template (Total tests = 148)	16425/16576	16404/16576	16419/16576
Overlapping Template	111/112	110/112	112/112
Universal	111/112	111/112	111/112
Approximate Entropy	110/112	112/112	112/112
Random Excursions (Total tests = 8)	603/608	572/576	561/568
Random Excursions Variant (Total tests = 18)	1352/1368	1291/1296	1276/1278
Serial (Total tests = 2)	222/224	222/224	221/224
Linear Complexity	112/112	112/112	112/112
Average Success Rate	**99.02%**	**99.05%**	**99.13%**

In contrast, for the gamma band, the average success rate is the highest at 99.05%, and it passes all of the NIST tests because all of the failure and no-uniformity rates are the lowest at 0%. Therefore, the gamma band is clearly random. These results approach the best results from other existing RNGs such as Blum-Blum-Shub [8], and Micali Schnorr [23] as shown in Table 11.

8 Conclusion

In this paper, we have proposed the new method to transform EEG signal and its wavebands into sequences of bits that can be used as a random number generators. Since our method does not require the use of seed to generate random numbers, it can be considered as TRNGs. It could also be used as a seed provider to improve randomness in PRNGs. Experimental results on the five datasets show that the proposed method achieves significantly high performance in success rate, and perform competitively in efficiency. The method will open potential possibilities of generating true random sequences of bits for biometrics-based systems to be added to the traditional ones based on physical systems.

For future work, this approach will be investigated further to validate the efficiency of our proposed method on large publicly available EEG datasets. Other biometric signals such as EMG (electromyogram) and ECG (electro-cardiography) will also be considered for randomness. Further candidates to biometric random number generators include blood volume pulse and similar easy-to-get measurements that, even when regular on the surface, may contain randomness in their internal structure. Any of these possibilities might be very important from the point of view of the implementation of our algorithms for generating random bit sequences via biometric-based methods.

References

[1] Acharya, U. R., Sree, S. V., Ang, P. C. A., Yanti, R., and Suri, J. S. (2012). Application of non-linear and wavelet based features for the automated identification of epileptic EEG signals. *Int. J. Neur. Syst.* 22, 1250002.

[2] Adeli, H., Ghosh-Dastidar, S., and Dadmehr, N. (2007). A wavelet-chaos methodology for analysis of EEGs and EEG subbands to detect seizure and epilepsy. IEEE Transactions on Biomedical Engineering, 54, 205–211. doi: 10.1109/TBME.2006.886855

[3] Al-Fahoum, A. S., and Al-Fraihat, A. A. (2014). Methods of EEG signal features extraction using linear analysis in frequency and time-frequency domains. *ISRN Neuroscience.* http://dx.doi.org/10.1155/2014/730218

[4] Allison, B. (2011). Trends in BCI research: progress today, backlash tomorrow?. XRDS: Crossroads, *The ACM Magazine for Students*, 18, 18–22. doi: 10.1145/2000775.2000784

[5] Anderson, C. W., Stolz, E. A., and Shamsunder, S. (1998). Multi-variate autoregressive models for classification of spontaneous elec-troencephalographic signals during mental tasks. *IEEE Transactions on Biomedical Engineering*, 45, 277–286. doi: 10.1109/10.661153

[6] Begleiter, H. (1999). EEG alcoholism database. Available at: https://kdd.ics.uci.edu/databases/eeg/eeg.data.html

[7] Bennett, C. H., and Brassard, G. (2014). Quantum cryptography: Public key distribution and coin tossing. *Theor. Comput. Sci.* 560, 7–11.

[8] Blum, L., Blum, M., and Shub, M. (1986). A simple unpredictable pseudo-random number generator. *SIAM J. Computing*, 15, 364–383..

[9] Brunner, C., Leeb, R., Müller-Putz, G., Schlögl, A., and Pfurtscheller, G. (2008). BCI Competition 2008–Graz data set A. *Institute for Knowledge*

Discovery (Laboratory of Brain-Computer Interfaces), Graz University of Technology, 136–142.

[10] Chen, G. (2014). Are electroencephalogram (EEG) signals pseudo-random number generators?. *J. Comput Appl. Math.* 268, 1–4.

[11] Click, T. H., Liu, A., and Kaminski, G. A. (2011). Quality of random number generators significantly affects results of Monte Carlo simulations for organic and biological systems. *J. Comput. Chem.* 32, 513–524.

[12] Dorrendorf, L., Gutterman, Z., and Pinkas, B. (2009). Cryptanalysis of the random number generator of the windows operating system. *ACM Transactions on Information and System Security (TISSEC),* 13, 10.

[13] Ferrenberg, A. M., Landau, D. P., and Wong, Y. J. (1992). Monte carlo simulations: Hidden errors from "good" random number generators. *Phy. Rev. Lett.* 69, 3382.

[14] Garrett, D., Peterson, D. A., Anderson, C. W., and Thaut, M. H. (2003). Comparison of linear, non-linear, and feature selection methods for EEG signal classification. *IEEE Transactions on Neural Systems and Rehabilitation Engineering,* 11, 141–144.

[15] Gerguri, S. (2008). *Biometrics Used for Random Number Generation* (Doctoral dissertation, Masarykova univerzita, Fakulta informatiky).

[16] Hartung, D., Wold, K., Graffi, K., and Petrovic, S. (2011). Towards a biometric random number generator – a general approach for true random extraction from biometric samples.

[17] Hunter, M. R. L. L., Smith, R. L., Hyslop, W., Rosso, O. A., Gerlach, R., Rostas, J. A. P., and Henskens, F. et al. (2005). The australian eeg database. Clinical EEG and neuroscience, 36(2), 76–81.

[18] Jonsson, P. (2011). Boom in Internet gambling ahead? US policy reversal clears the way.

[19] Koelstra, S., Muhl, C., Soleymani, M., Lee, J. S., Yazdani, A., Ebrahimi, T., and Patras, I. et al. (2012). Deap: A database for emotion analysis; using physiological signals. *IEEE Transactions on Affective Computing,* 3, 18–31.

[20] Marcel, S., and Millán, J. D. R. (2007). Person authentication using brainwaves (EEG) and maximum a posteriori model adaptation. *IEEE Transactions on Pattern Analysis and Machine Intelligence* 29, 743–752.

[21] Marton, K., and Suciu, A. (2015). On the interpretation of results from the NIST statistical test suite. *Romanian J. Inf. Sci. Technol.* 18, 18–32. http://www.imt.ro/romjist/Volum18/Number18_1/pdf/02-MSys.pdf

[22] Matyáš, V., and Říha, Z. (2010). Security of biometric authentication systems. In *International Conference on Computer Information Systems and Industrial Management Applications (CISIM)*, pp. 19–28. IEEE.

[23] Micali, S. and Schnorr, C. (1990) Pseudo-Random Sequence Generator, July 24 1990. US Patent 4,944,009.

[24] Ochoa, J. B. (2002). Eeg signal classification for brain computer interface applications. *Ecole Polytechnique Federale De Lausanne*, 7, 1–72. Available at: http://citeseerx.ist.psu.edu/viewdoc/download?doi=10.1.1.134.6148&rep=rep1&type=pdf

[25] O'Regan, S., Faul, S., and Marnane, W. (2010). Automatic detection of EEG artefacts arising from head movements. In *International Conference of Engineering in Medicine and Biology Society (EMBC)*, pp. 6353–6356. IEEE.

[26] Petchlert, B., and Hasegawa, H. (2014). Using a low-cost electroencephalogram (EEG) directly as random number generator. In *International Conference of Advanced Applied Informatics (IIAIAAI)*, pp. 470–474. IEEE.

[27] Petchlert, B., and Hasegawa, H. (2014). Using a low-cost electroencephalogram (EEG) directly as random number generator. In *International Conference of Advanced Applied Informatics (IIAIAAI)*, pp. 470–474. IEEE.

[28] Pijn, J. P., Van Neerven, J., Noest, A., and da Silva, F. H. L. (1991). Chaos or noise in EEG signals; dependence on state and brain site. *Electroencephalography and clinical Neurophysiology*, 79, 371–381.

[29] Rodríguez-Bermúdez, G., and García-Laencina, P. J. (2015).Analysis of eeg signals using non-linear dynamics and chaos: a review. *Appl. Math. Inf. Sci.* 9, 2309.

[30] Rukhin, A., Soto, J., Nechvatal, J., Barker, E., Leigh, S., Levenson, M., and Smid, M. et al. (2010). Statistical test suite for random and pseudorandom number generators for cryptographic applications, *NIST* special publication.

[31] Sanei, S., and Chambers, J. A. (2013). *EEG Signal Processing*. John Wiley & Sons.

[32] Sidorenko, A., and Schoenmakers, B. (2005). State Recovery Attacks on Pseudorandom Generators. *WEWoRC*, pp. 53–63.

[33] Stam, C. J. (2005). Non-Linear dynamical analysis of EEG and MEG: review of an emerging field. *Clin. Neurophysiology* 116, 2266–2301.

[34] Stipcevic, M. (2014). Preventing detector blinding attack and other random number generator attacks on quantum cryptography by use of an explicit random number generator. *arXiv preprint arXiv:1403.0143*.

[35] Szczepanski, J., Wajnryb, E., Amigó, J. M., Sanchez-Vives, M. V., and Slater, M. (2004). Biometric random number generators. *Comput. Security* 23, 77–84.

[36] Tong, S., and Thakor, N. V. (2009). *Quantitative EEG Analysis Methods and Clinical Applications*. Artech House.

[37] Vespa, P. M., Nenov, V., and Nuwer, M. R. (1999). Continuous EEG monitoring in the intensive care unit: early findings and clinical efficacy. *J. Clin. Neurophysiology* 16, 1–13.

[38] Wright, J. J., Kydd, R. R., and Liley, D. T. J. (1993). EEG models: Chaotic and linear. *Psycoloquy*, 4.

[39] Ying, L., Shu, W., Jing, Y., and Xiao, L. (2010, December). Design of a Random Number Generator from Fingerprint. In *International Conference of Computational and Information Sciences (ICCIS)*, pp. 278–280. IEEE.

Biographies

Dang van Nguyen is a Ph.D. student at the University of Canberra in Canberra, Australia since 2015. He attended Ha Noi University of Science, Viet Nam where he received his B.Sc. in Mathematics in 2005, and his M.Sc. in Mathematics in 2007. Dang has held a researcher position in Academy of Cryptography Technique – Vietnam Government Information Security Commission (VGISC), Ha Noi, Viet Nam since 2006. He has acquired a solid experience in binary sequence generators and cryptographic key generation methods. His Ph.D. research work focuses on EEG analysis for random number generators, and develops an EEG-based cryptographic key generation system for cryptography application.

Dat Tran received his B.Sc. and M.Sc. degrees from University of Science, Vietnam, in 1984 and 1994, respectively. He received his Graduate Diploma in Information Sciences and Ph.D. degree in Information Sciences & Engineering from University of Canberra, Australia, in 1996 and 2001, respectively. Currently, he is an Associate Professor at Faculty of Education Science, Technology and Mathematics, University of Canberra, Australia. His research areas include biometric authentication, security, pattern recognition and machine learning.

Wanli Ma received Ph.D. degree in September 2001. He was working at the Computer Services Center, University of Canberra, as an IT support officer. He became a lecturer at the School of Information Sciences and Engineering in January, 2004. Currently, he is Associate Dean Education of Faculty of Science and Technology at University of Canberra. His research areas include security, pattern recognition and machine learning.

Dharmendra Sharma is currently the Chair of University Academic Board and Professor of Computer Science at the University of Canberra (UC). He had been the Dean of the Faculty of Information Sciences and Engineering from 2007–2012 and as Head of School of the School of Information Sciences and Engineering from 2004–2007 at UC. Prof Sharma's research background is in the Artificial Intelligence areas of Planning, Data Analytics and Knowledge Discovery, Predictive Modelling, Constraint Processing, Fuzzy Reasoning, Brain-Computer Interaction, Hybrid Systems and their applications to health, education, security, digital forensics and sports.

Secure Data Sharing in Cloud Using an Efficient Inner-Product Proxy Re-Encryption Scheme

Masoomeh Sepehri[1], Alberto Trombetta[2] and Maryam Sepehri[1]

[1]*Department of Computer Science, University of Milan, Milan, Italy*
[2]*Department of Computer Science and Communication, University of Insubria, Varese, Italy*
E-mail: masoomeh.sepehri@unimi.it; maryam.sepehri@unimi.it; alberto.trombetta@uninsubria.it

Received 1 December 2017; Accepted 6 December 2017;
Publication 9 January 2018

Abstract

With the ever-growing production of data coming from multiple, scattered, highly dynamical sources (like those found in *IoT* scenarios), many providers are motivated to upload their data to the cloud servers and share them with other persons with different purposes. However, storing data on cloud imposes serious concerns in terms of data confidentiality and access control. These concerns get more attention when data is required to be shared among multiple users with different access policies. In order to update access policy without making re-encryption, we propose an efficient inner-product proxy re-encryption scheme that provides a proxy server with a transformation key with which a delegator's ciphertext associated with an attribute vector can be transformed to a new ciphertext associated with delegatee's attribute vector set. Our proposed policy updating scheme enables the delegatee to decrypt the shared data with its own key without requesting a new decryption key. We experimentally analyze the efficiency of our scheme and show that our scheme is adaptive attribute-secure against chosen-plaintext under standard Decisional Linear (*D-Linear*) assumption.

Keywords: Attribute-based cryptography, Secure data sharing, Fine-grained access control, Proxy re-encryption.

Journal of Cyber Security, Vol. 6_3, 339–378.
doi: 10.13052/jcsm2245-1439.635

1 Introduction

The emerging trend of sharing information among different users (esp. businesses and organizations) aiming to gain profit, has recently attracted a tremendous amount of attention from both research and industry communities. However, despite all benefits that data sharing inevitably provides [33], many organizations are reluctant to share their data with others due to the large initial investments of expensive infrastructure setup, large equipment, and daily maintenance cost [12]. With the advent of cloud computing; data outsourcing paradigm makes shared data much more accessible as users can retrieve them from anywhere with significant cost benefits. There are major concerns, with data confidentiality in the cloud as organizations lose control of their data and disclose sensitive information to a service provider that is not fully trusted. In addition, most organizations do not wish to grant full access privilege to other users. To this purpose, many research efforts have been dedicated to solve these issues by proposing cryptographically enforced access control mechanisms to set access policies for encrypted data such that only users with appropriate authorization can have access. Hence, many cryptographic-based approaches have been proposed and among them, attribute-based encryption (*ABE*) schemes [28] look very promising since they bind fine-grained access control policies to the data and they do not require an access control manager to check the access policies in real time. In *ABE* scheme, data is encrypted based on the set of attributes (key-policy *ABE* [10]) or according to an access control policy over attributes (ciphertext-policy *ABE* [2]), such that the decryption of ciphertext is possible only if a set of attributes in the user's private key matches with the attributes of the ciphertext, so that the data can be encrypted without exact knowledge of the users set that will be able to decrypt. Moreover, in *ABE* scheme senders and recipients are decoupled because they do not need to pre-share secrets, which simplifies key management for large-scale and dynamic systems and which makes data distribution more flexible. Furthermore, the *ABE* scheme is more strongly resistant to collusion attacks than traditional public key encryption schemes [2]. Although access control mechanisms based on *ABE* schemes present advantages regarding reduced communication, storage management and provide a fine-grained access control, they are not suitable for scenarios in which data must be shared among different parties with different access policies. A straightforward solution for applying a new access policy to the data is to decrypt the data and then re-encrypt it with a new access policy. However, this approach is very time-consuming and causes much computational overhead. These issues can be addressed by the adopting an attribute-based proxy re-encryption (*ABPRE*) scheme that delegates the re-encryption capability to a semi-trusted proxy who can transform the encrypted data to those encrypted under a different access policy by using the re-encryption key, which reduces the computational overhead of the data owner and the sensitive information as well as the user's private key cannot be revealed to the proxy. Although *ABPRE* approaches preserve the privacy of shared data among users with different access policies, they do not sufficiently protect the attributes associated with the ciphertexts. For example, in a healthcare scenario medical data require a high degree of privacy since they are accessed by many parties such as patients or staffs (e.g. doctors, nurses, care practitioners, etc.,) from a different department or belonging to different hospitals. Therefore, even partial exposure of those attributes could hurt the patient's privacy. Thus, access control system based on *ABE* are not enough to provide appropriate protection for sensitive data in some scenario like healthcare. Instead, predicate encryption (*PE*) scheme [14] can solve the above problems by offering the "attribute-hiding" property (which means that is not possible to determine the set of attributes with which the ciphertext is encrypted) as well as the "payload-hiding" property where a ciphertext conceal the plaintext. Informally, in the attribute-hiding, the secrecy of challenge attributes $\vec{x}_0$ and $\vec{x}_1$ is ensured against the adversary having private key $\vec{v}$ if the compatibility condition $< \vec{x}_0, \vec{v} > = < \vec{x}_1, \vec{v} >$ holds (here, $< \vec{x}, \vec{v} >$ denotes the standard inner product). In this work, we construct an adaptive secure attribute-hiding scheme through a proxy re-encryption method obtained with a non-trivial modification of a well known inner-product-based, attribute-based encryption scheme proposed by Park [26]. Unlike the existing scheme [32], we formally show that our proposal is adaptively secure for attribute-hiding (attribute-hiding in the sense of the definition by Katz *et al.* [14]).

The work is organized as follows: in Section 2 we present and analyze the related works; in Section 3 we introduce the definitions of inner-product encryption *IPE* and inner-product proxy re-encryption *IPPRE* and their security definitions; in Section 4 we present cryptographic primitives and complexity assumptions which are used in our proposed protocol; in Section 5 we provide a high-level view of the system in which we deploy in our scheme; in Section 6 we provide a detailed description of our proposed scheme and security proof based on standard game-based techniques; in Section 7 we show that our scheme can be efficiently implemented; finally, in Section 8 we draw some conclusions and propose some lines for future work.

2 Related Work

Proxy re-encryption scheme. Recently, several research papers have been developed for secure data sharing in the cloud [27, 30, 31]. Most of these works have adopted proxy re-encryption (*PRE*) scheme which was first proposed by Mambo and Okamoto [24] as a way to support the delegation of decryption rights. A seminal paper by Blaze *et al.* [3] proposed a bidirectional *PRE* scheme (called *BBS*) based on El-Gamal scheme [8] and introduced the notion of "*re-encryption key*". Using this key, a semi-trusted proxy server transforms a ciphertext encrypted under the delegator's public key into another ciphertext of the same plaintext encrypted under delegate's public key without revealing the underlying plaintext and user private key. Although *BBS* proxy re-encryption scheme is secure against chosen-plaintext attacks (*CPA*); however, it requires pre-sharing private key between parties in order to compute re-encryption key and has *bidirectional* property i.e., re-encryption key can be used to transform ciphertext from the delegator to the delegatee and vice versa, therefore it is only useful when the trust relationship between involved parties is mutual. Moreover, the scheme is not suitable for group communication since the proxy has to preserve n re-encryption key for n group members. Furthermore, *BBS* proxy re-encryption scheme exposed to collusion attacks, if the proxy colludes with one party they can recover the private key of the other party. To tackle these disadvantages, Ateniese *et al.* [1] proposed the first unidirectional and collusion resistant proxy re-encryption scheme without requiring pre-sharing between parties, based on bilinear maps.

Although proxy re-encryption techniques enable secure data sharing among different users in the cloud, they do not enforce fine-grained access control policies on the shared data. To address this issue, the traditional *PRE* approach has been extended with functionalities taken from attribute-based encryption (*ABE*) scheme in which both ciphertexts and user's private keys are associated with an attribute set and a user can decrypt a ciphertext only if the set of attributes in his private key match the attributes associated to the ciphertext [2, 10, 28].

Attribute-based proxy re-encryption scheme. An attribute-based proxy re-encryption (*ABPRE*) scheme was first proposed by Guo *et al.* [13] based on an (key-policy) attribute-based encryption scheme [10] and a general proxy re-encryption scheme. Under this scheme, a semi-trusted proxy server transforms a ciphertext associated with a set of attributes into a new ciphertext associated with different attributes set, without leakages about the plaintext and user private key. It has been proven that the security of the proposed scheme in the standard model based on decisional bilinear Diffie-Hellman (*DBDH*) assumption.

By adopting identity-based proxy re-encryption [11] to the construction of ciphertext-policy attribute-based encryption (*CP-ABE*) scheme [7], Liang *et al.* [21] presented the first ciphertext-policy attribute-based proxy re-encryption (*CP-ABPRE*) scheme. In this scheme, a proxy transforms a ciphertext generated under an access policy to another one corresponding to the same plaintext but to a different access policy. Their scheme satisfies *multi-hop* property and supports only access policies with *AND*-gates on positive and negative attributes (NOT). However, in this scheme, the size of the ciphertext increases linearly with the number of attributes in the system.

Later, Luo *et al.* [22] presented a ciphertext-policy *ABPRE*, which supports *AND*-gates access policies on multi-value attributes, negative attributes, and wildcards (which means the attributes don't appear in the *AND*-gates, therefore they are not considered in decryption algorithm). Their scheme satisfies the properties of *PRE*, such as *unidirectionality, non-interactivity* and *multi-hop*. Moreover, their scheme has two new properties: (i) *re-encryption control:* the encryptor can decide whether the ciphertext can be re-encrypted or not, and (ii) *extra access control:* the proxy can add extra access policy to the ciphertext during re-encryption process.

The computation cost of the previous *ABPRE* schemes is according to the number of attributes in the system, which implies huge computational overhead. To address this issue, based on Emura *et al.'s* [9] *CP-ABE* scheme which has a constant ciphertext length, Seo *et al.* [29] presented a *CP-ABPRE* scheme with constant number of pairing operations, which reduced significantly the computational cost and ciphertext length compared to previous *ABPRE* schemes. They reduced the number of pairing operation by using an exponential operation which can easily calculate the summation of the exponent. Therefore, they calculated the exponent and then computed the pairing operation just once. Their scheme can be adapted to various applications including e-mail forwarding and distributed file systems.

Most of the previous *CP-ABPRE* schemes [21, 22, 29, 36] only support *AND*-gates access structure on (multi-valued) positive and negative attributes. This limits their practical use. Therefore, it is desirable to propose a *CP-ABPRE* system supporting more expressive and flexible access policy. To tackle this issue, Li [18] presented a new *CP-ABPRE* scheme using matrix access structure policy which supports any monotonic access formula. Their scheme satisfies the properties of both *PRE* and *CP-ABPRE* schemes, such as *unidirectionality, non-interactivity, multi-hop, re-encryption control, extra access control* and *secret key security* providing a guarantee for the delegator such that if the proxy and all delegatees collude, they can not recover his master secret key. Moreover, they described the security model called Selective-Policy Model for their *CP-ABPRE* scheme based on [21].

The aforementioned *CP-ABPRE* schemes are only secure against selective chosen-plaintext attacks (*CPA*). The *CPA* security might not be sufficient enough in an open network since it only achieves the very basic requirement from an encryption scheme, which only allows an encryption to be secure against "passive" adversaries. Nevertheless, in a real network scenario, there might exist "active" adversaries trying to tamper an encryption in transit and next observing its decryption such that to obtain useful information related to the underlying data. Therefore, a *CP-ABPRE* system being secure against chosen-ciphertext attacks (*CCA*) is needed to prevent the above subtle attacks and enables the system to be further developed. To address this issue, based on the Waters's *CP-ABE* scheme [35], Liang *et al.* [19] proposed the first secure *CP-ABPRE* scheme against selective chosen-ciphertext attacks (*CCA*) which supports any monotonic access structures. Moreover, They constructed their proposed scheme in the random oracle model and they showed that their scheme can be proven *CCA* secure under the decisional q-parallel bilinear Diffie-Hellman exponent (q-parallel *BDHE*) assumption.

However, a *CP-ABPRE* system with selective security limits an adversary to choose an attack target before playing security game. Therefore, an adaptively *CCA* secure *CP-ABPRE* scheme is needed in most of the practical network applications. Thus, Liang *et al.* [20] proposed the first adaptively *CCA*-secure *CP-ABPRE* scheme by integrating the dual system encryption technology with selective proof technique. Their scheme supports any monotonic access structure such that users are allowed to fulfill more flexible delegation of decryption rights. This scheme is proven adaptively *CCA* secure in the standard model without loss of expressiveness on access policy. However, their scheme demands a number of paring operations that implies huge computational overheads.

Recently, Li *et al.* [17] proposed an efficient and adaptively secure *CP-ABPRE* scheme basing on Waters' dual system encryption technology [34]. This scheme is constructed in composite order bilinear groups and supports any monotone access structure. They proved that their scheme was secure under the complexity assumptions of the subgroup decision problem for 3 primes (*3P-SDP*). Compared with the existing schemes, their scheme requires a constant number of paring operations in Re-encryption and Decryption phases, which reduces the computational overhead.

Predicate encryption scheme. Although the attribute-based proxy re-encryption schemes have desirable functionality, they do not guarantee attribute-hiding property i.e., a ciphertext conceals the associated attributes as well as the plaintext so that no information about attributes is revealed during the decryption process. Therefore, to preserve the confidentiality of the attributes associated with the ciphertext, a seminal paper of Katz *et al.* [14] introduced the notion of predicate encryption (*PE*) as a generalized (fine-grained) notion of public key encryption that allows one to encrypt a message as well as attributes. In predicate encryption scheme, a ciphertext associated with attribute set $I \in \Sigma$ can be decrypted by a private key SK_f corresponding to the predicate $f \in \mathcal{F}$ if and only if $f(I) = True$. Katz *et al.* [14] also presented a special type of predicate encryption for a class of predicates called *inner-product encryption (IPE)*. In *IPE*, both ciphertext and private key are associated with vector $\overrightarrow{x}$ and $\overrightarrow{v}$ respectively and the ciphertext can be decrypted by the private key $SK_{\overrightarrow{v}}$ if and only if $< \overrightarrow{x}, \overrightarrow{v} >= 0$ (here, $< \overrightarrow{x}, \overrightarrow{v} >$ denotes the standard inner-product). Their method represents a wide class of predicates including conjunction and disjunction formulas and polynomial evaluations.

Later, Okamoto *et al.* [25] proposed a hierarchical predicate encryption (*HPE*) scheme for inner-product encryption. They used n-dimensional vector spaces in prime order bilinear groups and achieves full security under the standard model. In [16], Lewko *et al.* showed a fully secure *IPE* scheme based on composite bilinear groups resulting low practical efficiency. Although these *IPE* constructions achieve attribute-hiding properties, the security of their schemes is not under well-known standard assumptions. A different work of Park [26] presented an efficient *IPE* scheme supporting the attribute-hiding property. Their scheme is based on prime order bilinear groups and secure against the well-known Decision Bilinear Diffie-Hellman (*BDH*) and Decision Linear assumptions.

3 Definitions

In this section, we formally define the syntax of inner-product encryption (*IPE*) and inner-product proxy re-encryption (*IPPRE*) and their security properties. Our *IPE* definition follows the general framework of that given in [26]. Throughout this section, we consider the general case where Σ denotes an arbitrary set of attribute vectors and $\mathcal{F}$ denotes an arbitrary set of predicates involving inner-products over Σ.

Definition 1. An inner-product predicate encryption scheme (*IPE*) for the class of predicates $\mathcal{F}$ over the set of attributes Σ consists of PPT algorithms Setup, KeyGen, Encrypt and Decrypt such that:

Setup$_{\text{IPE}}$: takes as input a security parameter λ and a positive dimension n of vectors. It outputs a public key *PK* and a master secret key *MSK*.

KeyGen$_{\text{IPE}}$: takes as input a public key *PK,* a master secret key *MSK,* and a predicate vector $\overrightarrow{v} \in \mathcal{F}$. It outputs a private key $SK_{\overrightarrow{v}}$ associated with vector $\overrightarrow{v}$.

Encrypt$_{\text{IPE}}$: takes as input a public key *PK,* an attribute vector $\overrightarrow{x}$ and a message $M \in \mathcal{M}$. It outputs a corresponding ciphertext $CT_{\overrightarrow{x}} \leftarrow (PK, \overrightarrow{x}, M)$.

Decrypt$_{\text{IPE}}$: takes as input a private key $SK_{\overrightarrow{v}}$, and the ciphertext $CT_{\overrightarrow{x}}$. It outputs either a massage M if $f_{\overrightarrow{v}}(\overrightarrow{x}) = 1$, i.e., $< \overrightarrow{x}, \overrightarrow{v} > = 0$, or the distinguished symbol $\perp$ if $f_{\overrightarrow{v}}(\overrightarrow{x}) = 0$.

Definition 2. An inner-product predicate encryption scheme for predicate $\mathcal{F}$ over attributes Σ is attribute-hiding secure against adversary $\mathcal{A}$ under chosen plaintext attacks is given as follows:

Setup. The challenger runs the Setup algorithm and it gives the public key *PK* to the adversary $\mathcal{A}$.

Phase 1. The adversary $\mathcal{A}$ is allowed to adaptively issue a polynomial number of key queries. For a private key query $\overrightarrow{v}$, the challenger gives $SK_{\overrightarrow{v}}$ to $\mathcal{A}$.

Challenge. For a challenge query $(X_0, X_1, \overrightarrow{x}_0, \overrightarrow{x}_1)$, subject to the following restriction:

1. $< \overrightarrow{v}, \overrightarrow{x_0} > = < \overrightarrow{v}, \overrightarrow{x_1} > \neq 0$ for all private key queries $\overrightarrow{v}$, or
2. two challenge messages are equal, i.e $X_0 = X_1$, and any private key query $\overrightarrow{v}$ satisfies $< \overrightarrow{v}, \overrightarrow{x_0} > = < \overrightarrow{v}, \overrightarrow{x_1} >$.

The challenger flips a random $b \in \{0,1\}$ and computes the corresponding ciphertext as $CT_{\overrightarrow{x}_b} \leftarrow \text{Encrypt}(PK, \overrightarrow{x}_b, X_b)$. It then gives $CT_{\overrightarrow{x}_b}$ to the adversary.

Phase 2. The adversary $\mathcal{A}$ is allowed to adaptively issues polynomial number of key queries. For a private key query $\overrightarrow{v}$, subject to the aforementioned restrictions.

Finally, $\mathcal{A}$ outputs its guess $b' \in \{0,1\}$ for b and wins the game if $b = b'$. An advantage $\mathcal{A}$ in attacking *IPE* is defined as $Adv_{\mathcal{A}}^{\text{IPE-AH}}(\lambda) = Pr[b = b'] - \frac{1}{2}$. Therefore, an *IPE* scheme is *attribute-hiding* if all polynomial time adversaries have at most negligible advantage in the above game. If the restriction 1 in challenge is allowed for $\mathcal{A}$, an *IPE* scheme is *payload-hiding* if all polynomial time adversaries have at most negligible advantage in the game.

Definition 3. An inner-product proxy encryption (*IPPRE*) scheme creates a re-encryption key *ReKey* that gives the possibility of transforming a ciphertext associated with a vector $\overrightarrow{x}$ into a new ciphertext encrypting the same plaintext but associated with a different vector $\overrightarrow{w}$, while maintaining the confidentiality of the underlying plaintext. *IPPRE* scheme for the class of predicates $\mathcal{F}$ over n–dimensional vectors Σ for message space $\mathcal{M}$, consists of seven PPT algorithms Setup, Encrypt, KeyGen, Re-KeyGen, Re-Encrypt and Decrypt such that:

Setup$_{\text{IPPRE}}$: takes as input a security parameter λ and a dimension n of vectors. It outputs a public key *PK* and a master secret key *MSK*.

Encrypt$_{\text{IPPRE}}$: takes as input the public key *PK*, a vector $\overrightarrow{x} \in \Sigma$ of attributes and a message $M \in \mathcal{M}$ to output a ciphertext $CT_{\overrightarrow{x}}$.

KeyGen$_{\text{IPPRE}}$: takes as input the master secret key *MSK*, the public key *PK* and a predicate vector $\overrightarrow{v} \in \mathcal{F}$. It outputs a private key $SK_{\overrightarrow{v}}$ associated with vector $\overrightarrow{v}$.

Re-KeyGen$_{\text{IPPRE}}$: takes as input the master secret key *MSK* and two vectors $\overrightarrow{v}$ and $\overrightarrow{w}$. It outputs a re-encryption key $RK_{\overrightarrow{v},\overrightarrow{w}}$ that transforms a ciphertext that could be decrypted by $SK_{\overrightarrow{v}}$ into a ciphertext encrypted with vector $\overrightarrow{w}$.

Re-Encrypt$_{\text{IPPRE}}$: takes as input a re-encryption key $RK_{\overrightarrow{v},\overrightarrow{w}}$ and a ciphertext $CT_{\overrightarrow{x}}$ to output a re-encrypted ciphertext $CT'_{\overrightarrow{x}}$.

Decrypt$_{\text{IPPRE}}$: takes as input the ciphertext $CT_{\overrightarrow{x}}$ and the private key $SK_{\overrightarrow{v}}$. It outputs either a message M if $f_{\overrightarrow{v}}(\overrightarrow{x}) = 1$, i.e., $< \overrightarrow{x}, \overrightarrow{v} > = 0$, or the distinguished symbol $\perp$ if $f_{\overrightarrow{v}}(\overrightarrow{x}) = 0$.

From here on, we use the terms *Level-1* (*L1*) and *Level-2* (*L2*) to denote ciphertexts obtained as the output of Encrypt and Re-Encrypt algorithms, respectively.

Correctness. The correctness property requires to decrypt the ciphertext by the appropriate private key. More precisely, for the two levels *L1* and *L2* we have:

L1: Decrypt (KeyGen ($MSK, PK, \overrightarrow{v}$), Encrypt ($PK, \overrightarrow{x}, M$)) $= M$;

L2: Decrypt (KeyGen (MSK, PK, v'), Re-Encrypt (Re-KeyGen (KeyGen ($MSK, PK, \overrightarrow{v}$), $\overrightarrow{w}$), $CT_{\overrightarrow{x}}$)), $= M$,

where $\overrightarrow{x}$ satisfies $\overrightarrow{v}$, $\overrightarrow{w}$ satisfies v', MSK is a master secret key, PK is a public key, $CT_{\overrightarrow{x}}$ is a ciphertext related to message M and an attribute vector $\overrightarrow{x}$.

Definition 4 (Attribute-Hiding for Level-1 Ciphertexts (*AH-L1*)). An inner-product proxy re-encryption (*IPPRE*) scheme, predicate $\mathcal{F}$ over vectors Σ is attribute-hiding secure *Level-1* against adversary $\mathcal{A}$ under chosen-plaintext attacks (*CPA*) if for all probabilistic polynomial-time *PPT*, the advantage of $\mathcal{A}$ in the following security game Γ is negligible in the security parameter.

Setup. The challenger $\mathcal{B}$ runs Setup (λ, n) algorithm and gives the public key PK to $\mathcal{A}$.

Phase 1. $\mathcal{A}$ adaptively makes a polynomial number of queries as:

(a) **Private key query:** For a private key query $\overrightarrow{v}$, the challenger gives $SK_{\overrightarrow{v}} \xleftarrow{R}$ KeyGen (MSK, PK, $\overrightarrow{v}$) to $\mathcal{A}$, where R indicates that $SK_{\overrightarrow{v}}$ is randomly selected from KeyGen according to its distribution.

(b) **Re-encryption key query:** For a re-encryption key query with ($\overrightarrow{v}$, $\overrightarrow{w}$), the challenger computes $RK_{\overrightarrow{v},\overrightarrow{w}} \xleftarrow{R}$ Re-KeyGen (MSK, $\overrightarrow{v}$, $\overrightarrow{w}$) where $SK_{\overrightarrow{v}} \xleftarrow{R}$ KeyGen (MSK, PK, $\overrightarrow{v}$) and gives the re-encryption key to the adversary.

(c) **Re-encryption query:** For a re-encryption query ($\overrightarrow{v}$, $\overrightarrow{w}$, $CT_{\overrightarrow{x}}$), $\mathcal{B}$ computes the re-encryption key $RK_{\overrightarrow{v},\overrightarrow{w}} \xleftarrow{R}$ Re-KeyGen (MSK, $\overrightarrow{v}$, $\overrightarrow{w}$), where $SK_{\overrightarrow{v}} \xleftarrow{R}$ KeyGen (MSK, PK, $\overrightarrow{v}$) and $CT'_{\overrightarrow{x}} \xleftarrow{R}$ Re-Encrypt (PK, $RK_{\overrightarrow{v},\overrightarrow{w}}$, $CT_{\overrightarrow{x}}$, $\overrightarrow{w}$).

Challenge. For a challenge query $(\overrightarrow{x}_0, \overrightarrow{x}_1, M_0, M_1)$ under the condition that:

    – Any private key query $\overrightarrow{v}$ and re-encryption key query $(\overrightarrow{v}_l, \overrightarrow{w}_l)$, for $l = 1, \ldots, p_1$ where p_1 is the maximum number of private key queries requested by the adversary, $M_0 = M_1$ if $< \overrightarrow{v}, \overrightarrow{x}_0 >=< \overrightarrow{v}, \overrightarrow{x}_1 >= 0$ and $< \overrightarrow{v}_l, \overrightarrow{x}_0 >=< \overrightarrow{v}_l, \overrightarrow{x}_1 >= 0$ in the case that $< \overrightarrow{v}, \overrightarrow{w}_l >= 0$.

The challenger $\mathcal{B}$ samples a random bit $b \overset{U}{\leftarrow} \{0, 1\}$, where U indicates that b is uniformly selected from $\{0,1\}$ and gives $CT_{\overrightarrow{x}_b} \overset{R}{\leftarrow}$ Encrypt $(PK, \overrightarrow{x}_b, M_b)$ to $\mathcal{A}$.

Phase 2. $\mathcal{A}$ may continue to request private key queries, re-encryption key queries and re-encryption queries subject to the same restrictions as before and the condition for the re-encryption queries.

    Re-encryption Query: For a re-encryption query of the form $(\overrightarrow{v}_t, \overrightarrow{w}_t, CT_t)$, for $t = 1, \ldots, p_2$ where p_2 is the maximum number of re-encrypted queries, under the condition that $M_0 = M_1$ if $< \overrightarrow{v}_t, \overrightarrow{x}_0 >=< \overrightarrow{v}_t, \overrightarrow{x}_1 >= 0$ and $< \overrightarrow{v'}, \overrightarrow{w}_t >= 0$ for any decryption key query for $\overrightarrow{v'}$ if $CT_t = CT_{\overrightarrow{x}_b}$.

    The challenger computes $RK_{\overrightarrow{v}_t, \overrightarrow{w}_t} \overset{R}{\leftarrow}$ Re-KeyGen $(MSK, \overrightarrow{v}_t, \overrightarrow{w}_t)$ and $CT'_{\overrightarrow{w}_t} \overset{R}{\leftarrow}$ Re-Encrypt $(PK, RK_{\overrightarrow{v}_t, \overrightarrow{w}_t}, CT_t)$, and it gives $CT'_{\overrightarrow{w}_t}$ to the adversary.

Guess. $\mathcal{A}$ outputs a bit b' and succeeds if $b' = b$.

    Hence, we define the advantage $\mathcal{A}$ as $Adv_{\mathcal{A}}^{\text{AH-L1}}(\lambda) := \Pr[b = b'] - \frac{1}{2}$. The *IPPRE* scheme is attribute-hiding *Level-1* ciphertext if all polynomial time adversaries have at most negligible advantage in the above game.

Definition 5 (Attribute-Hiding for Level-2 Re-encrypted Ciphertexts (*AH-L2*)). An inner-product proxy re-encryption (*IPPRE*) scheme, predicate $\mathcal{F}$ over vectors Σ is attribute-hiding secure *Level-2* against adversary $\mathcal{A}$ under chosen-plaintext attacks (*CPA*) if for all probabilistic polynomial-time PPT, the advantage of $\mathcal{A}$ in the following security game Γ is negligible in the security parameter.

Setup, Phase 1: These algorithms are defined as the same as those we defined in Definition 4, respectively.

Challenge: Upon receiving the query $(\overrightarrow{x}_0, \overrightarrow{x}_1, M_0, M_1, \overrightarrow{v}_0, \overrightarrow{v}_1, \overrightarrow{w}_0, \overrightarrow{w}_1)$ from the adversary with the restrictions that $(M_0, \overrightarrow{x}_0, \overrightarrow{v}_0) = (M_1, \overrightarrow{x}_1, \overrightarrow{v}_1)$

if $< \vec{v'}, \vec{w}_0 >=< \vec{v'}, \vec{w}_1 >= 0$, for any private key query $\vec{v'}$, the challenger $\mathcal{B}$ samples a random bit $b \xleftarrow{R} \{0, 1\}$ and gives:

$CT'_{\vec{w}_b} \xleftarrow{R}$ Re-Encrypt(PK, Re-KeyGen(PK, KeyGen(PK, SK, $\vec{v}_b$), $\vec{w}_b$), Encrypt $(PK, \vec{x}_b, M_b))$. Then the challenger gives the result to the adversary.

Phase 2: The adversary $\mathcal{A}$ may continue to request private key queries, re-encryption key queries and re-encryption queries under the restrictions we mentioned in challenge phase.

Guess: $\mathcal{A}$ outputs its guess $b' \in \{0, 1\}$ b and wins the game $b = b'$.

We define the advantage of $\mathcal{A}$ as $\text{Adv}_{\mathcal{A}}^{\text{PAH-L2}}(\lambda) := \Pr[b = b'] - \frac{1}{2}$. Hence, the scheme is predicate- and attribute-hiding for re-encrypted ciphertexts if all polynomial time adversaries have at most negligible advantage in the above game. To prove this statement for each run of the game, we define a variable $s_{M, \vec{x}, \vec{v}} := 0$ if $(M_0, \vec{x}_0, \vec{v}_0) \neq (M_1, \vec{x}_1, \vec{v}_1)$ for challenge $(M_l, \vec{x}_l, \vec{v}_l)$ for $l = 0, 1$ and $s_{M, \vec{x}, \vec{v}} := 1$, otherwise.

4 Cryptographic Background and Complexity Assumptions

In this section, we define Bilinear Map following the notation in [5] and review some general assumptions we use in Section 6 to prove the security of our construction.

Bilinear Map. Let $\mathbb{G}$ and $\mathbb{G}_T$ be two multiplicative cyclic groups of prime order p, and g be a generator of $\mathbb{G}$. A pairing (or bilinear map) $e : \mathbb{G} \times \mathbb{G} \to \mathbb{G}_T$ is a function that has the following properties [5]:

1. *Bilinear:* a map $e : \mathbb{G} \times \mathbb{G} \to \mathbb{G}_T$ is bilinear if $e(u^a, v^b) = e(u, v)^{ab}$ for all $u, v \in \mathbb{G}$ and $a, b \in \mathbb{Z}_p^*$.
2. *Non-degenerate:* $e(g, g) \neq 1$. The map does not send all pairs in $\mathbb{G} \times \mathbb{G}$ to the identity in $\mathbb{G}_T$. Since $\mathbb{G}$ and $\mathbb{G}_T$ are groups of prime order, this implies that if g is a generator of $\mathbb{G}$ then $e(g, g)$ is a generator of $\mathbb{G}_T$.
3. *Computable:* there is an efficient algorithm to compute the map $e(u, v)$ for any $u, v \in \mathbb{G}$.

A map e is an admissible bilinear map in $\mathbb{G}$ if satisfies the three properties above. Note that $e(\ ,\ )$ is symmetric since $e(g^a, g^b) = e(g, g)^{ab} = e(g^b, g^a)$.

Decisional Bilinear Diffie-Hellman (DBDH) Assumption [5]. Let $a,b,c \in \mathbb{Z}_p^*$ be chosen at random and g be a generator for $\mathbb{G}$. The Decisional *BDH* assumption is defined as follows: given $(g, g^a, g^b, g^c, Z) \in \mathbb{G}^4 \times \mathbb{G}_T$ as input, determine whether $Z = e(g, g)^{abc}$ or Z is a random in $\mathbb{G}_T$.

The Decision Linear (*D-Linear*) Assumption [4]. Let $z_1, z_2, z_3, z_4, \in \mathbb{Z}_p^*$ be chosen at random and g be a generator for $\mathbb{G}$. The Decision Linear assumption is defined as follows: given $(g, g^{z_1}, g^{z_2}, g^{z_1 z_3}, g^{z_2 z_4}, Z) \in \mathbb{G}^6$ as input, determine whether $Z = g^{z_3 + z_4}$ or Z is random in $\mathbb{G}$. We consider an equivalently modified version such as: given $(g, g^{z_1}, g^{z_2}, g^{z_1 z_3} g^{z_4}, Z) \in \mathbb{G}^6$ as input, determine whether $Z = g^{z_2(z_3 + z_4)}$ or Z is random in $\mathbb{G}$.

Definition 6. We say that the {Decision *BDH*, Decision Linear} assumption holds in $\mathbb{G}$ if the advantage of any polynomial time algorithm is solving the {Decision *BDH*, Decision Linear} problem is negligible.

5 System Model

In this section, we present a streamlined version of secure and private data sharing system in a healthcare environment to show how to deploy an *IPPRE* scheme in such real-world scenarios. A *IPPRE*-based data sharing healthcare system including five entities *Data Owner, Authorized Users Owner, Cloud Storage Server, Trust Authority* and the *Proxy Server* works as follows:

Initialization. This step is run by a *Trust Authority* (*TA*) who is responsible for key issuing and attribute management. As shown in Figure 1, the authority first generates master secret key *MSK* and public key *PK*, and then distributes *PK* and access policy A_i to each data owner i (e.g., Owner 1 from hospital 1 and Owner 2 from hospital 2). It also generates private keys for *Authorized Users* (User 1 (e.g., a group of care practitioners) and User 2 (e.g., specialist)).

Data Upload. This step is run at data owner side. Consider that the owner 1 from hospital 1 is willing to store and share its medical records via the *Cloud Storage Server* in such a way that only care practitioners from hospital 1 can have access. The owner 1 encrypts its own data (e.g., message M_1) under a set of associated attributes A_1 (e.g., Encrypt$_{A_1}$ (M_1)), where A_1 indicates access privilege on the owner 1's data. In a similar way, the owner 2 uploads its encrypted message (M_2).

Data Access. This step is run between the authorized users and the cloud server (*Level-1*) or between authorized users through the cloud using a proxy server (*Level-2*).

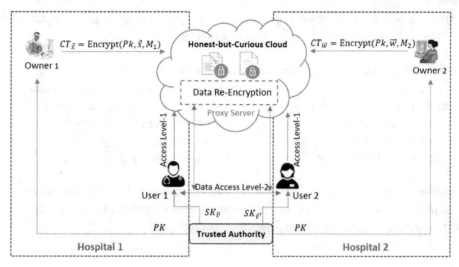

Figure 1   *IPPRE*-based data sharing healthcare system model.

1. *Level-1*. User 1 who satisfies A_1 can access to the owner 1's data using its own private key associated with a vector $\overrightarrow{v}$.

2. *Level-2*. There are some situations in which the user 1 needs to share the owner 1's medical data with the user 2 from hospital 2 who is able to decrypt only the ciphertexts associated with an access policy A_2 (attribute vector $\overrightarrow{w}$ in Figure 1), but not the access policy A_1 (attribute vector $\overrightarrow{x}$ in Figure 1). In this case, a *Proxy Server* is used to translate the data encrypted with access policy A_1 to the one under access policy A_2 in an efficient way without revealing the data (payload-hiding property) and its corresponding attributes (attribute-hiding property).

In our system model, we assume that the cloud and proxy server are honest-but-curious i.e., they will correctly execute the protocol, and will not deny services to the authorized users. But they are curious to learn information about data contents.

6 The Main Construction

In this section, we construct our *IPPRE* scheme in detail and give intuition about our proof. The scheme consists of six algorithms namely Setup, Encrypt, KeyGen, Decrypt, Re-KeyGen, Re-Encrypt. We describe our construction with considering the following assumptions:

Assumptions:

- some positive integer n, $\Sigma = (\mathbb{Z}_p^*)^n$ is the set of attributes,
- a vector $\overrightarrow{v} = (v_1, \ldots, v_n) \in \Sigma$, each component v_i belong to the set $\mathbb{Z}_p^*$, and
- a message $M \in \mathcal{M}$ and a vector $\overrightarrow{v}$, each $\overrightarrow{v}$ belongs to Σ and $\mathcal{M} = \mathbb{G}_T$.

6.1 The Construction

(PK, MSK) $\leftarrow$ **Setup** (λ, n). On input a security parameter $\lambda \in \mathbb{Z}^+$ and the number of attributes n, Setup algorithm runs *init*$(\lambda)^1$ to get the tuple $(p, \mathbb{G}, \mathbb{G}_T, e)$. It then picks a random generator $g \in \mathbb{G}$, random exponents $\delta_1, \delta_2, \theta_1, \theta_2, \{w_{1,i}\}_{i=1}^n, \{t_{1,i}\}_{i=1}^n, \{f_{1,i}, f_{2,i}\}_{i=1}^n, \{h_{1,i}, h_{2,i}\}_{i=1}^n$ in $\mathbb{Z}_p^*$. It also picks a random $g_2 \in \mathbb{G}$ and a random $\Omega \in \mathbb{Z}_p^*$ to obtain $\{w_{2,i}\}_{i=1}^n, \{t_{2,i}\}_{i=1}^n$ in $\mathbb{Z}_p^*$ under constraints that:

$$\Omega = \delta_1 w_{2,i} - \delta_2 w_{1,i}, \qquad \Omega = \theta_1 t_{2,i} - \theta_2 t_{1,i}.$$

For $i = 1, \ldots, n$, the Setup algorithm first computes:

$$W_{1,i} = g^{w_{1,i}}, \quad W_{2,i} = g^{w_{2,i}}, \quad T_{1,i} = g^{t_{1,i}}, \quad T_{2,i} = g^{t_{2,i}},$$
$$F_{1,i} = g^{f_{1,i}}, \quad F_{2,i} = g^{f_{2,i}}, \quad H_{1,i} = g^{h_{1,i}}, \quad H_{2,i} = g^{h_{2,i}},$$

and then sets:

$$U_1 = g^{\delta_1}, \quad U_2 = g^{\delta_2}, \quad V_1 = g^{\theta_1}, \quad V_2 = g^{\theta_2}, \quad g_1 = g^{\Omega}, \Lambda = e(g, g_2).$$

Finally, the Setup algorithm outputs the public key *PK* (including $(p, \mathbb{G}, \mathbb{G}_T, e)$) and master secret key *MSK* as:

$$PK = (g, g_1, \{W_{1,i}, W_{2,i}, F_{1,i}, F_{2,i}\}_{i=1}^n, \{T_{1,i}, T_{2,i}, H_{1,i}, H_{2,i}\}_{i=1}^n,$$
$$\{U_i, V_i\}_{i=1}^2, \Lambda) \in \mathbb{G}^{8n+6} \times \mathbb{G}_T$$
$$MSK = (\{w_{1,i}, w_{2,i}, t_{1,i}, t_{2,i}, f_{1,i}, f_{2,i}, h_{1,i}, h_{2,i}\}_{i=1}^n, \{\delta_i, \theta_i\}_{i=1}^2, g_2)$$
$$\in \mathbb{Z}_p^{8n+4} \times \mathbb{G}.$$

---

[1] *init* is an algorithm that takes as input a security parameter 1^n and outputs a tuple $(p, \mathbb{G}, \mathbb{G}_T, e)$, where $\mathbb{G}$ and $\mathbb{G}_T$ are groups of prime order p and $e : \mathbb{G} \times \mathbb{G} \rightarrow \mathbb{G}_T$ is a bilinear map.

$SK_{\overrightarrow{v}} \leftarrow$ **KeyGen** $(MSK, PK, \overrightarrow{v})$. On input vector $\overrightarrow{v} = (v_1, \ldots, v_n) \in (\mathbb{Z}_p^*)^n$, public key PK and master secret key MSK, the algorithm randomly picks exponents $\lambda_1, \lambda_2, \{r_i\}_{i=1}^n, \{\phi_i\}_{i=1}^n$ in $\mathbb{Z}_p^*$ to output the private key as:

$$SK_{\overrightarrow{v}} = (K_A, K_B, \{K_{1,i}, K_{2,i}\}_{i=1}^n, \{K_{3,i}, K_{4,i}\}_{i=1}^n) \in \mathbb{G}^{4n+2} \text{ where :}$$

$$\{K_{1,i} = g^{-\delta_2 r_i} g^{\lambda_1 v_i w_{2,i}}, \quad K_{2,i} = g^{\delta_1 r_i} g^{-\lambda_1 v_i w_{1,i}}\}_{i=1}^n,$$

$$\{K_{3,i} = g^{-\theta_2 \phi_i} g^{\lambda_2 v_i t_{2,i}}, \quad K_{4,i} = g^{\theta_1 \phi_i} g^{-\lambda_2 v_i t_{1,i}}\}_{i=1}^n,$$

$$K_A = g_2 \prod_{i=1}^n K_{1,i}^{-f_{1,i}} K_{2,i}^{-f_{2,i}} K_{3,i}^{-h_{1,i}} K_{4,i}^{-h_{2,i}}, K_B = \prod_{i=1}^n g^{-(r_i+\phi_i)}.$$

$CT_{\overrightarrow{x}} \leftarrow$ **Encrypt** $(PK, \overrightarrow{x}, M)$. On input vector $\overrightarrow{x} = (x_1, \ldots, x_n) \in (\mathbb{Z}_p^*)^n$, a message $M \in \mathbb{G}_T$ and the public key PK, the algorithm selects random exponents $s_1, s_2, s_3, s_4 \in \mathbb{Z}_p^*$ to get ciphertext $CT_{\overrightarrow{x}}$ as the follows:

$$CT_{\overrightarrow{x}} = (A, B, \{C_{1,i} = W_{1,i}^{s_1} \cdot F_{1,i}^{s_2} \cdot U_1^{x_i s_3}, C_{2,i} = W_{2,i}^{s_1} \cdot F_{2,i}^{s_2} \cdot U_2^{x_i s_3}\}_{i=1}^n,$$

$$\{C_{3,i} = T_{1,i}^{s_1} \cdot H_{1,i}^{s_2} \cdot V_1^{x_i s_4}, C_{4,i} = T_{2,i}^{s_1} \cdot H_{2,i}^{s_2} \cdot V_2^{x_i s_4}\}_{i=1}^n, \Lambda^{-s_2} M) \in \mathbb{G}^{4n+2}$$

$$\times \mathbb{G}_T.$$

Where we define each component of $CT_{\overrightarrow{x}}$ as the following, $1 \le i \le n$:

$$A = g^{s_2}, B = g_1^{s_1} = g^{s_1 \Omega}, \quad D = \Lambda^{-s_2} M$$

$$C_{1,i} = g^{w_{1,i} s_1} g^{f_{1,i} s_2} g^{\delta_1 x_i s_3}, \quad C_{2,i} = g^{w_{2,i} s_1} g^{f_{2,i} s_2} g^{\delta_2 x_i s_3}$$

$$C_{3,i} = g^{t_{1,i} s_1} g^{h_{1,i} s_2} g^{\theta_1 x_i s_4}, \quad C_{4,i} = g^{t_{2,i} s_1} g^{h_{2,i} s_2} g^{\theta_2 x_i s_4}.$$

In this step, random elements $\{W_{1,i}^{s_1}, W_{2,i}^{s_1}, T_{1,i}^{s_1}, T_{2,i}^{s_1}, \}$ are used to mask each component x_i of a vector $\overrightarrow{x}$. For instance, the ciphertext $C_{1,i}$ is in the form $W_{1,i}^{s_1} F_{1,i}^{s_2} U_1^{x_i s_3}$, which is not easily tested even if we use prime order groups equipped with a symmetric bilinear map. If we omit $W_{1,i}^{s_1}$, the resulting term $F_{1,i}^{s_2} U_1^{x_i s_3}$ is enough for hiding x_i component, however, for the case that $x_i = 0$ in $\mathbb{Z}_p^*$, the term becomes $F_{2,i}^{s_2}$ that can be tested as $e(A, F_{1,i}) \overset{?}{=} e(g, C_{1,i})$ using bilinear maps.

$RK_{\overrightarrow{v}, \overrightarrow{w}} \leftarrow$ **Re-KeyGen** $(MSK, \overrightarrow{v}, \overrightarrow{w})$. The algorithm first calls the KeyGen algorithm and picks a random $d \in \mathbb{Z}_p^*$ to compute g_2^d and $g_2^{d\delta_2}, g_2^{-d\delta_1}, g_2^{d\theta_2}, g_2^{-d\theta_1}$. It then calls the Encrypt algorithm to encrypt g_2^d under the vector $\overrightarrow{w}$ using Encrypt $(PK, \overrightarrow{w}, g_2^d)$ and outputs $CT_{\overrightarrow{w}}$.

To compute the re-encryption key, the Re-KeyGen algorithm picks random exponents $\lambda_1', \lambda_2', \{r_i'\}_{i=1}^n, \{\phi_i'\}_{i=1}^n$ in $\mathbb{Z}_P^*$ and computes $RK_{\overrightarrow{v},\overrightarrow{w}}$, $1 \leq i \leq n$ as:

$$K_A' = g_2 \prod_{i=1}^n K_{1,i}'^{-f_{1,i}} K_{2,i}'^{-f_{2,i}} K_{3,i}'^{-h_{1,i}} K_{4,i}'^{-h_{2,i}}, \quad K_B' = \prod_{i=1}^n g^{-(r_i'+\phi_i')}.$$

Where we have:

$$K_{1,i}' = g^{-\delta_2 r_i'} g^{\lambda_1' v_i w_{2,i}} g_2^{d\delta_2}, \quad K_{2,i}' = g^{\delta_1 r_i'} g^{-\lambda_1' v_i w_{1,i}} g_2^{-d\delta_1}$$

$$K_{3,i}' = g^{-\theta_2 \phi_i'} g^{\lambda_2' v_i t_{2,i}} g_2^{d\theta_2}, \quad K_{4,i}' = g^{\theta_1 \phi_i'} g^{-\lambda_2' v_i t_{1,i}} g_2^{-d\theta_1}.$$

The Re-KeyGen algorithm with the inputs vectors $\overrightarrow{v}$, $\overrightarrow{w}$ consists of two parts: a modified decryption key vector $\overrightarrow{v}$ and a ciphertext encrypted with vector $\overrightarrow{w}$. The modified decryption key differs from a normal decryption key: in the decryption procedure, a normal decryption key combines with elements of the ciphertext to recover the binding factor that is used for hiding the message (e.g., $e(g, g_2)^{-s_2}$); the modified decryption key instead produces the product of the blinding factor with another new binding factor. This new blinding factor can only be removed with the combination of a group element encrypted in the Re-KeyGen algorithm (e.g., g_2^d) and the element $B = g^{s_1\Omega}$ in the ciphertext. Therefore, the *Level-2* access of the Decrypt algorithm consists of the original blinded message, the product with the new blinding factor obtained by decrypting the original ciphertext with the modified decryption key in the Re-Encrypt algorithm, the element B from the original ciphertext and the ciphertext component of the Re-KeyGen algorithm.

$CT_{\overrightarrow{x}}' \leftarrow$ **Re-Encrypt** $(RK_{\overrightarrow{v},\overrightarrow{w}}, CT_{\overrightarrow{x}})$. On input a re-encryption key $RK_{\overrightarrow{v},\overrightarrow{w}}$ and $CT_{\overrightarrow{x}}$ ciphertext, the algorithm outputs $CT_{\overrightarrow{x}}' = (A, B, CT_{\overrightarrow{w}}, \hat{CT}, D)$

$$A = g^{s_2}, B = g_1^{s_1} = g^{s_1\Omega}, D = \Lambda^{-s_2} M, CT_{\overrightarrow{w}} = \text{Encrypt}(PK, \overrightarrow{w}, g_2^d)$$

computing $\hat{CT}$, the algorithm checks if the attributes list in $RK_{\overrightarrow{v},\overrightarrow{w}}$ satisfies the attributes set of $CT_{\overrightarrow{x}}$ if not, returns $\perp$; otherwise, $1 \leq i \leq n$, it calculates the following pairings to output $\hat{CT}$:

$$\prod_{i=1}^n e(C_{1,i}, K_{1,i}') \cdot e(C_{2,i}, K_{2,i}') \cdot e(C_{3,i}, K_{3,i}') \cdot e(C_{4,i}, K_{4,i}')$$

Where we have:

$$e(C_{1,i}, K'_{1,i}) = e(g^{w_{1,i}s_1} g^{f_{1,i}s_2} g^{\delta_1 x_i s_3}, g^{-\delta_2 r'_i} g^{\lambda'_1 v_i w_{2,i}} g_2^{d\delta_2})$$

$$e(C_{2,i}, K'_{2,i}) = e(g^{t_{1,i}s_1} g^{h_{2,i}s_2} g^{\theta_2 x_i s_4}, g^{\theta_1 \phi'_i} g^{-\lambda'_1 v_i t_{1,i}} g_2^{-d\theta_1})$$

$$e(C_{3,i}, K'_{3,i}) = e(g^{t_{2,i}s_1} g^{h_{2,i}s_2} g^{\theta_2 x_i s_4}, g^{\theta_1 \phi'_i} g^{-\lambda'_1 v_i t_{1,i}} g_2^{-d\theta_1})$$

$$e(C_{4,i}, K'_{4,i}) = e(g^{t_{2,i}s_1} g^{h_{2,i}s_2} g^{\theta_2 x_i s_4}, g^{\theta_1 \phi'_i} g^{-\lambda'_1 v_i t_{1,i}} g_2^{-d\theta_1}).$$

Hence, we expand the above formula as follows:

$$\prod_{i=1}^{n} e(C_{1,i}, K'_{1,i}) \cdot e(C_{2,i}, K'_{2,i}) \cdot e(C_{3,i}, K'_{3,i}) \cdot e(C_{4,i}, K'_{4,i})$$

$$= \prod_{i=1}^{n} e(g^{w_{1,i}s_1} g^{f_{1,i}s_2} g^{\delta_1 x_i s_3}, g^{-\delta_2 r'_i} g^{\lambda'_1 v_i w_{2,i}} g_2^{d\delta_2})$$

$$\cdot e(g^{w_{2,i}s_1} g^{f_{2,i}s_2} g^{\delta_2 x_i s_3}, g^{\delta_1 r'_i} g^{-\lambda'_1 v_i w_{1,i}} g_2^{-d\delta_1})$$

$$\cdot e(g^{t_{1,i}s_1} g^{h_{1,i}s_2} g^{\theta_1 x_i s_4}, g^{-\theta_2 \phi'_i} g^{\lambda'_1 v_i t_{2,i}} g_2^{d\theta_2})$$

$$\cdot e(g^{t_{2,i}s_1} g^{h_{2,i}s_2} g^{\theta_2 x_i s_4}, g^{\theta_1 \phi'_i} g^{-\lambda'_1 v_i t_{1,i}} g_2^{-d\theta_1})$$

$$= \prod_{i=1}^{n} e(g^{w_{1,i}s_1}, g^{-\delta_2 r'_i}) \cdot e(g^{f_{1,i}s_2}, g^{-\delta_2 r'_i} g^{\lambda'_1 v_i w_{2,i}} g_2^{d\delta_2})$$

$$\cdot e(g^{\delta_1 x_i s_3}, g^{\lambda'_1 v_i w_{2,i}}) \cdot e(g^{w_{1,i}s_1}, g_2^{d\delta_2}) \cdot e(g^{w_{2,i}s_1}, g^{\delta_1 r'_i})$$

$$\cdot e(g^{f_{2,i}s_2}, g^{\delta_1 r'_i} g^{-\lambda'_1 v_i w_{1,i}} g_2^{-d\delta_1}) \cdot e(g^{\delta_2 x_i s_3}, g^{-\lambda'_1 v_i w_{1,i}}) \cdot e(g^{w_{2,i}s_1}, g_2^{-d\delta_1})$$

$$\cdot e(g^{t_{1,i}s_1}, g^{-\theta_2 \phi'_i}) \cdot e(g^{h_{1,i}s_2}, g^{-\theta_2 \phi'_i} g^{\lambda'_2 v_i t_{2,i}} g_2^{d\theta_2}) \cdot e(g^{\theta_1 x_i s_4}, g^{\lambda'_2 v_i t_{2,i}})$$

$$\cdot e(g^{t_{1,i}s_1}, g_2^{d\theta_2}) \cdot e(g^{t_{2,i}s_1}, g^{\theta_1 \phi'_i}) \cdot e(g^{h_{2,i}s_2}, g^{\theta_1 \phi'_i} g^{-\lambda'_1 v_i t_{1,i}} g_2^{-d\theta_1})$$

$$\cdot e(g^{\theta_2 x_i s_4}, g^{-\lambda'_1 v_i t_{1,i}}) \cdot e(g^{t_{2,i}s_1}, g_2^{-d\theta_1})$$

$$= \prod_{i=1}^{n} e(g^{-\delta_2 w_{1,i}}, g^{r_i' s_1}) \cdot e(g^{s_2}, (g^{-\delta_2 r_i'} g^{\lambda_1' v_i w_{2,i}} g_2^{d\delta_2})^{f_{1,i}})$$

$$\cdot e(g,g)^{\lambda_1' \delta_1 w_{2,i} x_i v_i s_3} \cdot e(g^{w_{1,i} s_1}, g_2^{d\delta_2}) \cdot e(g^{-\delta_1 w_{2,i}}, g^{r_i' s_1})$$

$$\cdot e(g^{s_2}, (g^{\delta_1 r_i'} g^{-\lambda_1' v_i w_{1,i}} g_2^{d\delta_1})^{f_{2,i}}) \cdot e(g,g)^{\lambda_1' \delta_2 w_{1,i} x_i v_i s_3}$$

$$\cdot e(g^{w_{2,i} s_1}, g_2^{-d\delta_1}) \cdot e(g^{-\theta_2 t_{1,i}}, g^{\phi_i' s_1}) \cdot e(g^{s_2}, (g^{-\theta_2 \phi_i'} g^{\lambda_1' v_i t_{2,i}} g_2^{d\theta_2})^{h_{1,i}})$$

$$\cdot e(g,g)^{\lambda_2' \theta_1 t_{2,i} x_i v_i s_4} \cdot e(g^{t_{1,i} s_1}, g_2^{d\theta_2}) \cdot e(g^{\theta_1 t_{2,i}}, g^{\phi_i' s_1})$$

$$\cdot e(g^{s_2}, (g^{\theta_1 \phi_i'} g^{-\lambda_1' v_i t_{1,i}} g_2^{-d\theta_1})^{h_{2,i}}) \cdot e(g,g)^{-\lambda_2' \theta_2 t_{1,i} x_i v_i s_4}$$

$$\cdot e(g^{t_{2,i} s_1}, g_2^{-d\theta_1}).$$

$$= \prod_{i=1}^{n} e(g^{\delta_1 w_{2,i} - \delta_2 w_{1,i}}, g^{r'_i s_1}) \cdot e(g^{\theta_1 t_{2,i} - \theta_2 t_{1,i}}, g^{\phi'_i s_1})$$

$$\cdot e(g^{s_2}, K_{1,i}'^{f_{1,i}} K_{2,i}'^{f_{2,i}} K_{3,i}'^{h_{1,i}} K_{4,i}'^{h_{2,i}})$$

$$\cdot e(g,g)^{[\lambda_1'(\delta_1 w_{2,i} - \delta_2 w_{1,i}) s_3 + \lambda_2'(\theta_1 t_{2,i} - \theta_2 t_{1,i}) s_4] x_i v_i}$$

$$\cdot e(g^{-\delta_1 w_{2,i} + \delta_2 w_{1,i}}, g_2^{ds_1}) \cdot e(g^{-\theta_1 t_{2,i} + \theta_2 t_{1,i}}, g_2^{ds_1})$$

$$= e(g^{\Omega s_1}, \prod_{i=1}^{n} g^{(r_i' + \phi_i')}) \cdot e(g^{s_2}, \prod_{i=1}^{n} K_{1,i}'^{f_{1,i}} K_{2,i}'^{f_{2,i}} K_{3,i}'^{h_{1,i}} K_{4,i}'^{h_{2,i}})$$

$$\cdot e(g,g)^{\Omega(\lambda_1' s_3 + \lambda_2' s_4) < \vec{x}, \vec{v} >} \cdot e(g^{-\Omega}, g_2^{ds_1}).$$

Finally, the algorithm outputs $\hat{C}T$ to obtain:

$$\hat{C}T = e(A, K_A') \cdot e(B, K_B') \cdot \prod_{i=1}^{n} e(C_{1,i}, K_{1,i}') \cdot e(C_{2,i}, K_{2,i}')$$

$$\cdot e(C_{3,i}, K_{3,i}') \cdot e(C_{4,i}, K_{4,i}')$$

$$= e(g^{s_2}g_2 \prod_{i=1}^{n} K'_{1,i}{}^{-f_{1,i}} K'_{2,i}{}^{-f_{2,i}} K'_{3,i}{}^{-h_{1,i}} K'_{4,i}{}^{-h_{2,i}}) \cdot e(g^{\Omega s_1}, \prod_{i=1}^{n} g^{-(r'_i+\phi'_i)})$$

$$\cdot e(g^{\Omega s_1}, \prod_{i=1}^{n} g^{(r'_i+\phi'_i)}) \cdot e(g^{s_2}, \prod_{i=1}^{n} K'_{1,i}{}^{f_{1,i}} K'_{2,i}{}^{f_{2,i}} K'_{3,i}{}^{h_{1,i}} K'_{4,i}{}^{h_{2,i}})$$

$$\cdot e(g^{-\Omega}, g_2^{ds_1}) \cdot e(g,g)^{\Omega(\lambda'_1 s_3 + \lambda'_2 s_4)<\vec{x},\vec{v}>}$$

$$= e(g^{s_2}, g_2) \cdot e(g,g)^{\Omega(\lambda'_1 s_3 + \lambda'_2 s_4)<\vec{x},\vec{v}>} \cdot e(g^{-\Omega}, g_2^{ds_1}).$$

$M \leftarrow$ **Decrypt** $(CT_{\vec{x}}, SK_{\vec{v}})$. On input the ciphertext $CT_{\vec{x}}$ and a private key $SK_{\vec{v}}$, the algorithm proceeds differently according to two *Level-1* or *Level-2* access:

1. **Level-1 access.** If $CT_{\vec{x}}$ is an original well-formed ciphertext, then algorithm decrypts $CT_{\vec{x}} = (A, B, \{C_{1,i}, C_{2,i}\}_{i=1}^n, \{C_{3,i}, C_{4,i}\}_{i=1}^n, D = e(g,g_2)^{-s_2}M)$ using the private key $SK_{\vec{v}}(K_A, K_B\{K_{1,i}, K_{2,i}\}_{i=1}^n, \{K_{3,i}, K_{4,i}\}_{i=1}^n)$ to output message M:

$$M \leftarrow D \cdot e(A, K_A) \cdot e(B, K_B)$$

$$\cdot \prod_{i=1}^{n} e(C_{1,i}, K_{1,i}) \cdot e(C_{2,i}, K_{2,i}) \cdot e(C_{3,i}, K_{3,i}) \cdot e(C_{4,i}, K_{4,i}).$$

In this step, the masking elements used in Encrypt algorithm have to be canceled out. To this purpose, the proposed scheme generates two relative pairing values, a positive and a negative in order to be removed at the end. This can be checked by the following equality:

$$e(C_{1,i}, K_{1,i}) \cdot e(C_{2,i}, K_{2,i}) = e(g^{w_{1,i}s_1}g^{f_{1,i}s_2}g^{\delta_1 x_i s_3}, g^{-\delta_2 r_i}g^{\lambda_1 v_i w_{2,i}})$$

$$\cdot e(g^{w_{2,i}s_1}g^{f_{2,i}s_2}g^{\delta_2 x_i s_3}, g^{\delta_1 r_i}g^{-\lambda_1 v_i w_{1,i}})$$

where both $e(g^{w_{1,i}s_1}, g^{\lambda_1 v_i w_{2,i}})$ and $e(g^{\delta_1 x_i s_3}, g^{-\delta_2 r_i})$ are canceled out. Additionally, we need to remove $e(g^{w_{1,i}s_1}, g^{-\delta_2 r_i}) \cdot e(g^{w_{2,i}s_1}, g^{\delta_1 r_i})$ that are changed into one pairing as $e(g^{\Omega s_1}, g^{r_i})$. This value is also eliminated by the additional computation of $e(B, K_B)$ in the decryption procedure.

Correctness.

Assume the ciphertext $CT_{\vec{x}}$ is well-formed the vector $\vec{x} = x_1 \ldots, x_n$. Then, we have:

$$D \cdot e(A, K_A) \cdot e(B, K_B) \cdot \prod_{i=1}^{n} e(C_{1,i}, K_{1,i}) \cdot e(C_{2,i}, K_{2,i}) \cdot e(C_{3,i}, K_{3,i})$$

$$\cdot e(C_{4,i}, K_{4,i})$$

$$= e(g, g_2)^{-s_2} M \cdot e(g, g_2)^{s_2} \cdot e(g, g)^{\Omega(\lambda_1 s_3 + \lambda_2 s_4) < \vec{x}, \vec{v} >}$$

$$= M \cdot e(g, g)^{\Omega(\lambda_1 s_3 + \lambda_2 s_4) < \vec{x}, \vec{v} >}.$$

It is worth noting that the term $e(g, g_2)^{s_2}$ is generated from the pairing computation $e(A, K_A) = e(g_2 \prod_{i=1}^{n} K_{1,i}^{-f_{1,i}} K_{2,i}^{-f_{2,i}} K_{3,i}^{-h_{1,i}} K_{4,i}^{-h_{2,i}})$. Thus, the output of the above result is M if $< \vec{x}, \vec{v} >= 0$ in $\mathbb{Z}_p^*$. If $< \vec{x}, \vec{v} >\neq 0$ in $\mathbb{Z}_p^*$, then there is only such case that $\lambda_1 s_3 + \lambda_2 s_4 = 0$ in $\mathbb{Z}_p^*$ with probability at most $1/p$, as in the predicate-only *IPE* scheme.

2. **Level-2 access** (from here on referred to as Re-Decrypt). If $CT_{\vec{x}}$ is a re-encrypted well-formed ciphertext, then it is of the form $CT_{\vec{x}}' = (A, B, CT_{\vec{w}}, \hat{CT}, D = e(g, g_2)^{-s_2} M)$. The algorithm first decrypts $CT_{\vec{w}}$ using $(SK_{\vec{v}'}$ as above to obtain g_2^d as Decrypt $(SK_{\vec{v}'}, CT_{\vec{w}}) \to g_2^d$.

Then, it calculates: $\bar{CT} = e(B, g_2^d) = e(g^{s_1 \Omega}, g_2^d)$ and obtains the message as $M \leftarrow D \cdot \hat{CT} \cdot \bar{CT}$

The *Level-2* access of the Decrypt algorithm consists of the original blinded message, the product with the new blinding factor obtained by decrypting the original ciphertext with the modified decryption key in the Re-Encrypt algorithm, the element B from the original ciphertext and the ciphertext component of the Re-KeyGen algorithm. To decrypt a re-encrypted ciphertext of *Level-2* access, the proposed scheme first decrypts the ciphertext component of the Re-KeyGen algorithm to obtain the group element, then combines this group element with the element B from the original ciphertext to use the result removing both the original blinding factor of the message and the new binding factor introduced by the Re-Encrypt algorithm. Finally, the message is recovered if the vector $\vec{x}$ associated with the ciphertext and the vector $\vec{v}$ associated with the private key m orthogonal vectors (e.g., $< \vec{x}, \vec{v} >= 0$).

Correctness.

To verify the correctness, we compute $D \cdot \hat{CT} \cdot \bar{CT}$ as:

$$e(g, g_2)^{-s_2} M \cdot e(g^{s_2}, g_2) \cdot e(g, g)^{\Omega(\lambda_1' s_3 + \lambda_1' s_4) < \overrightarrow{x}, \overrightarrow{v} >} \cdot e(g^{-\Omega}, g_2^{ds_1})$$

$$= e(g, g_2)^{-s_2} M \cdot e(g, g_2)^{s_2} \cdot e(g, g)^{\Omega(\lambda_1' s_3 + \lambda_2' s_4) < \overrightarrow{x}, \overrightarrow{v} >}$$

$$\cdot e(g, g_2)^{-s_1 \Omega d} \cdot e(g, g_2)^{s_1 \Omega d}$$

$$= M \cdot e(g, g)^{\Omega(\lambda_1' s_3 + \lambda_2' s_4) < \overrightarrow{x}, \overrightarrow{v} >}.$$

The result outputs M if $< \overrightarrow{x}, \overrightarrow{v} > = 0$ in $\mathbb{Z}_p^*$. If $< \overrightarrow{x}, \overrightarrow{v} > \neq 0$ in $\mathbb{Z}_p^*$, then there is only such case that $(\lambda_1' s_3 + \lambda_2' s_4) = 0$ in $\mathbb{Z}_p^*$ with probability at most $1/p$, as in the predicate-only *IPE* scheme.

Level-1 access. If $CT_{\overrightarrow{x}}$ is an original well-med ciphertext, then algorithm decrypts $CT_{\overrightarrow{x}} = (A, B, \{C_{1,i}, C_{2,i}\}_{i=1}^n, \{C_{3,i}, C_{4,i}\}_{i=1}^n, D = e(g, g_2)^{-s_2} M)$ using the private key:

$SK_{\overrightarrow{v}} = (K_A, K_B, \{K_{1,i}, K_{2,i}\}_{i=1}^n, \{K_{3,i}, K_{4,i}\}_{i=1}^n)$ to output message M:

$$M \leftarrow D \cdot e(A, K_A) \cdot e(B, K_B)$$

$$\cdot \prod_{1=1}^n e(C_{1,i}, K_{1,i}) \cdot e(C_{2,i}, K_{2,i}) \cdot e(C_{3,i}, K_{3,i}) \cdot e(C_{4,i}, K_{4,i}).$$

In this step, the masking elements used in Encrypt algorithm have to be canceled out. To this purpose, the proposed scheme generates two relative pairing values, a positive and a negative in order to be removed at the end. This can be checked by the following equality:

$$e(C_{1,i}, K_{1,i}) \cdot e(C_{2,i}, K_{2,i}) = e(g^{w_{1,i} s_1} g^{f_{1,i} s_2} g^{\delta_1 x_i s_3}, \delta^{-\delta_2 r_i} g^{\lambda_1 v_i w_{2,i}})$$

$$\cdot e(g^{w_{2,i} s_1} g^{f_{2,i} s_2} g^{\delta_2 x_i s_3}, g^{\delta_1 r_i} g^{-\lambda_1 v_i w_{1,i}}),$$

where both $e(g^{w_{1,i} s_1}, g^{\lambda_1 v_i w_{2,i}})$ and $e(g^{\delta_1, i x_i s_3}, g^{-\delta_2 r_i})$ are canceled out. Additionally, we need to remove $e(g^{w_{1,i} s_1}, g^{-\delta_2 r_i}) \cdot e(g^{w_{2,i} s_1}, g^{\delta_1 r_i})$ that are changed into one pairing as $e(g^{\Omega s_1}, g^{r_i})$. This value is also eliminated by the additional computation of $e(B, K_B)$ in the decryption procedure.

Correctness.

Assume the ciphertext $CT_{\overrightarrow{x}}$ is well-formed the vector $\overrightarrow{x} = x_1, \ldots, x_n$. Then, we have:

$$D \cdot e(A, K_A) \cdot e(B, K_B) \cdot \prod_{i=1}^{n} e(C_{1,i}, K_{1,i}) \cdot e(C_{2,i}, K_{2,i}) \cdot e(C_{3,i}, K_{3,i})$$

$$\cdot e(C_{4,i}, K_{4,i})$$

$$= e(g, g_2)^{-s_2} M \cdot e(g, g_2)^{s_2} \cdot e(g, g)^{\Omega(\lambda_1 s_3 + \lambda_2 s_4) < \overrightarrow{x}, \overrightarrow{v} >}$$

$$= M \cdot e(g, g)^{\Omega(\lambda_1 s_3 + \lambda_2 s_4) < \overrightarrow{x}, \overrightarrow{v} >}$$

It is worth noting that the term $e(g, g_2)^{s_2}$ is generated from the pairing computation of $e(A, K_A) = e(g_2 \prod_{i=1}^{n} K_{1,i}{}^{-f_{1,i}}{}_{2,i}{}^{-f_{2,i}} K_{3,i}{}^{-h_{1,i}} K_{4,i}{}^{-h_{2,i}})$. Thus, the output of the above result is M if $< \overrightarrow{x}, \overrightarrow{v} > = 0$ in $\mathbb{Z}_p^*$. If $< \overrightarrow{x}, \overrightarrow{v} > \neq 0$ in $\mathbb{Z}_p^*$, then there is only such case that $\lambda_1 s_3 + \lambda_2 s_4 = 0$ in $\mathbb{Z}_p^*$ with probability at most $1/p$, as in the predicate-only *IPE* scheme.

Level-2 access (from here on referred to as Re-Decrypt). If $CT_{\overrightarrow{x}}$ is a re-encrypted well-formed ciphertext, then it is of the form $CT'_{\overrightarrow{x}} = (A, B, CT_{\overrightarrow{w}}, \hat{CT}, D = e(g, g_2)^{-s_2}M)$. The algorithm first decrypts $CT_{\overrightarrow{w}}$ using $SK_{\overrightarrow{v}'}$ as above to obtain $g_2{}^d$ as Decrypt $(SK_{\overrightarrow{v}'}, CT_{\overrightarrow{w}}) \rightarrow g_2{}^d$.

Then, it calculates: $\bar{CT} = e(B, g_2{}^d) = e(g^{s_1 \Omega}, g_2{}^d)$ and obtains the message as $M \leftarrow D \cdot \hat{CT} \cdot \bar{CT}$.

The *Level-2* access of the Decrypt algorithm consists of the original blinded message, the product with the new blinding factor obtained by decrypting the original ciphertext with the modified decryption key in the Re-Encrypt algorithm, the element B from the original ciphertext and the ciphertext component of the Re-KeyGen algorithm. To decrypt a re-encrypted ciphertext of *Level-2* access, the proposed scheme first decrypts the ciphertext component of the Re-KeyGen algorithm to obtain the group element, then combines this group element with the element B from the original ciphertext to use the result removing both the original blinding factor of the message and the new binding factor introduced by the Re-Encrypt algorithm. Finally, the message is recovered if the vector $\overrightarrow{x}$ associated with the ciphertext and the vector $\overrightarrow{v}$ associated with the private key m orthogonal vectors (e.g., $< \overrightarrow{x}, \overrightarrow{v} > = 0$).

Correctness.

To verify the correctness, we compute $D \cdot \hat{CT} \cdot \bar{CT}$ as:

$$e(g, g_2)^{-s_2} M \cdot e(g^{s_2}, g_2) \cdot e(g, g)^{\Omega(\lambda_1' s_3 + \lambda_2' s_4) < \overrightarrow{x}, \overrightarrow{y} >} \cdot e(g^{-\Omega}, g_2^{ds_1})$$

$$\cdot e(g^{s_1 \Omega}, g_2^d) = e(g, g_2)^{-s_2} M \cdot e(g, g_2)^{s_2} \cdot e(g, g)^{\Omega(\lambda_1' s_3 + \lambda_2' s_4) < \overrightarrow{x}, \overrightarrow{y} >}$$

$$\cdot e(g, g_2)^{-s_1 \Omega d} \cdot e(g, g_2)^{s_1 \Omega d} = M \cdot e(g, g)^{\Omega(\lambda_1' s_3 + \lambda_2' s_4) < \overrightarrow{x}, \overrightarrow{y} >}.$$

The result outputs M if $< \overrightarrow{x}, \overrightarrow{v} >= 0$ in $\mathbb{Z}_p^*$. If $< \overrightarrow{x}, \overrightarrow{v} >\neq 0$ in $\mathbb{Z}_p^*$, then there is only such case that $(\lambda_1' s_3 + \lambda_2' s_4)$ in $\mathbb{Z}_p^*$ with probability at most $1/p$, as in the predicate-only *IPE* scheme.

6.2 Proof of Security

Here, we describe a mechanism to show that our proposed scheme achieves the security requirements according to the definitions stated in the Section 3. For *Level-1* and *Level-2* ciphertext challenge, an adversary may request private key, re-encryption key and re-encryption queries by choosing vectors $(\overrightarrow{x}_0, \overrightarrow{x}_1, \overrightarrow{w}_0, \overrightarrow{w}_1)$ at the beginning of the security game. For instance, in the case of *Level-1* access, the adversary outputs two vectors $\overrightarrow{x}_0, \overrightarrow{x}_1$ and queries corresponding to a vector $\overrightarrow{v}$ such that $< \overrightarrow{v}, \overrightarrow{x}_0 >=< \overrightarrow{v}, \overrightarrow{x}_1 >= 0$ where $M_0 = M_1$. The adversary goal is to decide which one of the two vectors is associated with the challenge ciphertext. In the case of *Level-2* access, the adversary outputs challenge vectors $\overrightarrow{x}_0, \overrightarrow{x}_1$ along with $\overrightarrow{w}_0, \overrightarrow{w}_1$ for re-encryption keys. The adversary goal is to decide which one of the two vectors $\overrightarrow{w}_0, \overrightarrow{w}_1$ is associated with the re-encrypted query.

To prove the *Level-1* access, similarly to [14] we suppose that our encryption system contains two parallel sub-systems. That is, a challenge ciphertext will be encrypted with respect to one vector in the first subsystem and a different vector in a second sub-system. Let $(\overrightarrow{a}, \overrightarrow{b})$ denote a ciphertext encrypted using $\vec{0}$ vector (that is orthogonal to everything) in intermediate game to prove indistinguishably when encrypting to $\overrightarrow{x}_0$ corresponding to $(\overrightarrow{x}_0, \overrightarrow{x}_0)$ and when encrypting to $\overrightarrow{x}_1$ corresponding to $(\overrightarrow{x}_1, \overrightarrow{x}_1)$ as:

$$(\overrightarrow{x}_0, \overrightarrow{x}_0) \approx (\overrightarrow{x}_0, \overrightarrow{0}) \approx (\overrightarrow{x}_0, \overrightarrow{x}_1) \approx (\overrightarrow{0}, \overrightarrow{x}_1) \approx (\overrightarrow{x}_1, \overrightarrow{x}_1).$$

This structure allows us to use a simulator (challenger) that will essentially work in one subsystem without knowing what is happening in the other one [14]. It determines whether a sub-system encrypts the given vector or the zero vector. Details of this proof is given in [26].

To prove the *Level-2* access, we apply game transformation proof [15] with a multiple sequence of games whose aim are to change components of the challenge ciphertext to independent ones from challenge bit b (random form). In the following we discuss it in details.

In the following, we show that the proposed *IPPRE* scheme is predicate- and attribute-hiding re-encrypted ciphertext (*Level-2*) against chosen-plaintext attacks provided the underlying *IPE* scheme under the Decision Linear assumption holds in $\mathbb{G}$.

Proof of Theorem 1 (PAH-L2: Predicate- and Attribute-hiding Re-encrypted ciphertext)

We consider two cases in the proof of Theorem 1 according to the value of $s_{M,\vec{x},\vec{v}}$ mentioned in the Definition 5. This value holds the following claims:

- For any private key query $\vec{v'}$, the variable $s_{M,\vec{x},\vec{v}} = 0$ when it holds $< \vec{v'}, \vec{w}_0 >=< \vec{v'}, \vec{w}_1 > \neq 0$.
- For any private key query $\vec{v'}$, the variable $s_{M,\vec{x},\vec{v}} = 1$ when it holds $< \vec{v'}, \vec{w}_0 >=< \vec{v'}, \vec{w}_1 >$.

Theorem 1. The *IPPRE* scheme is predicate- and attribute-hiding for re-encrypted ciphertexts against chosen-plaintext attacks provided underlying *IPE* scheme is fully attribute-hiding. For any adversary $\mathcal{A}$ there exist probabilistic machines $\epsilon_{1-1}, \epsilon_{1-2}, \epsilon_{2-1}$ and ϵ_{2-2} whose running times are essentially the same as that of $\mathcal{A}$, such that for any security parameter λ.

$$Adv_{\mathcal{A}}^{\text{PAH-L2}}(\lambda) \leq Adv_{\epsilon_{1-1}}^{\text{IPE-AH}}(\lambda) + Adv_{\epsilon_{1-2}}^{\text{IPE-AH}}(\lambda) + \frac{1}{2}(Adv_{\epsilon_{2-1}}^{\text{IPE-AH}}(\lambda)$$
$$+ Adv_{\epsilon_{2-2}}^{\text{IPE-AH}}(\lambda)).$$

Proof. We execute a preliminary game transformation from Game 0 (original game in Definition 5) to Game $0'$, which is the same as Game 0 except flip a coin $\tau_{M,\vec{x},\vec{v}}$ before setup, and the game is aborted at the final step if $\tau_{M,\vec{x},\vec{v}} \neq s_{M,\vec{x},\vec{v}}$. Hence, the advantage of Game $0'$ is a half of that in Game 0. The value $\tau_{M,\vec{x},\vec{v}}$ is chosen independently from $s_{M,\vec{x},\vec{v}}$, and therefore the probability that the game is aborted is $\frac{1}{2}$ that is $Adv_{\mathcal{A}}^{(0')}(\lambda) = \frac{1}{2} \cdot Adv_{\mathcal{A}}^{\text{PAH-L2}}(\lambda)$. Moreover, $\Pr[\mathcal{A} \ wins] = \frac{1}{2}(\Pr(\mathcal{A} \ wins | \tau_{M,\vec{x},\vec{v}} = 0) + (\Pr(\mathcal{A} \ wins | \tau_{M,\vec{x},\vec{v}} = 1))$ in Game $0'$.

Hence, we have:

$$Adv_{\mathcal{A}}^{\text{PAH-L2}}(\lambda) \leq Adv_{\epsilon_{1-1}}^{\text{IPE-AH}}(\lambda) + Adv_{\epsilon_{1-2}}^{\text{IPE-AH}}(\lambda) + \frac{1}{2}(Adv_{\epsilon_{2-1}}^{\text{IPE-AH}}(\lambda)$$
$$+ Adv_{\epsilon_{2-2}}^{\text{IPE-AH}}(\lambda)).$$

Therefore, to show how our scheme is predicate- and attribute-hiding for re-encrypted ciphertext under *D-Linear* assumption, we consider the two cases as bellow:

Proof of Theorem 1 in the case $\tau_{M,\vec{x},\vec{v}} = 0$

Lemma 1. The proposed *IPPRE* scheme is predicate- and attribute-hiding for re-encrypted ciphertexts against chosen-plaintext attack in the case $\tau_{M,\vec{x},\vec{v}}$ under the attribute hiding underlying *IPE* scheme.

For any adversary $\mathcal{A}$, there exists probabilistic mechanisms ϵ_{1-1} and ϵ_{1-2}, whose running times are essentially the same as that of $\mathcal{A}$ such that for any security parameter λ in the case $\tau_{M,\vec{x},\vec{v}} = 0$.

$$\Pr[\mathcal{A}_{wins}|\tau_{M,\vec{x},\vec{v}} = 0] - \frac{1}{2} \leq \text{Adv}_{\epsilon_{1-1}}^{IPE\text{-}AH} + \text{Adv}_{\epsilon_{1-2}}^{IPE\text{-}AH}.$$

The aim is that $CT_{\vec{w}}$ is changed to a ciphertext with random attribute and random attribute message. We apply the game transformation consisting of three games Game $0'$, Game 1 and Game 2. In Game 1, the $CT_{\vec{w}_b}$ under vector $\vec{w}_b$ is changed to $CT_{\vec{r}} = \text{Encrypt}\,(PK, \vec{r}, R)$ where $\vec{r}$ chosen uniformly random from Σ and random value $R \in \mathbb{G}_T$.

In the case $\tau_{M,\vec{x},\vec{v}} = 0$, the adversary does not request private key query $\vec{v}$ such that $< \vec{x}_b, \vec{v} >= 0$. Hence, $CT_{\vec{x}_b}$ is changed to $\text{Encrypt}_{IPE}\,(PK, \vec{r}, R)$ by using the attribute-hiding security underlying *IPE* scheme.

Proof of Lemma 1. In order to prove the Lemma 1, we consider the following games. We only describe the components which are changed in the other games.

Game $0'$. Same as Game 0 except that flip a coin $\tau_{M,\vec{x},\vec{v}} \xleftarrow{U} \{0,1\}$ before setup, and the game is aborted if $\tau_{M,\vec{x},\vec{v}} \neq S_{M,\vec{x},\vec{v}}$. We consider the case with $\tau_{M,\vec{x},\vec{v}} = 0$ and rely to the challenge query $(\vec{x}_0, \vec{x}_1, M_0, M_1, \vec{v}_0, \vec{v}_1, \vec{w}_0, \vec{w}_1)$ as the following:

$(\boldsymbol{A} = g^{s_2}, \boldsymbol{B} = g_1^{s_1} = g^{s_1\Omega}, \Lambda = e(g, g_2)^{-s_2},$

$CT_{\vec{x}_b} = \text{Encrypt}\,(PK, \vec{x}_b, M_b)$

$\quad = (g^{s_2}, g_1^{s_1}, \{W_{1,i}^{s_1} \cdot F_{1,i}^{s_2} \cdot U_1^{x_{ib}s_3}, W_{2,i}^{s_1} \cdot F_{2,i}^{s_2} \cdot U_2^{x_{ib}s_3}\}_{i=1}^n,$

$\quad \{T_{1,i}^{s_1} \cdot H_{1,i}^{s_2} \cdot V_1^{x_{ib}s_4}, T_{2,i}^{s_1} \cdot H_{2,i}^{s_2} \cdot V_2^{x_{ib}s_4}\}_{i=1}^n, \Lambda^{-s_2} M_b) \in \mathbb{G}^{4n+2} \times \mathbb{G}_T,$

$CT_{\vec{w}_b} = \text{Encrypt}\left(PK, \vec{w}_b, g_2^d\right)$

$\quad = (g^{s_2}, g_1^{s_1}, \{W_{1,i}^{s_1} \cdot F_{1,i}^{s_2} \cdot U_1^{w_{ib}s_3}, W_{2,i}^{s_1} \cdot F_{2,i}^{s_2} \cdot U_2^{w_{ib}s_3}\}_{i=1}^n,$

$\quad \left\{T_{1,i}^{s_1} \cdot H_{1,i}^{s_2} \cdot V_1^{w_{ib}s_4}, T_{2,i}^{s_1} \cdot H_{2,i}^{s_2} \cdot V_2^{w_{ib}s_4}\}_{i=1}^n, \Lambda^{-s_2} g_2^d\right) \in \mathbb{G}^{4n+2} \times \mathbb{G}_T,$

$$\hat{CT} = e(g^{s_2}, g_2) \cdot e(g,g)^{\Omega(\lambda_1' s_3 + \lambda_2' s_4) < \vec{x}_b, \vec{v}_b >} \cdot e(g^{-\Omega}, g_2^{ds_1})).$$

Game 1. Game 1 is the same as Game $0'$ except that the reply to challenge query for $(\overrightarrow{x}_0, \overrightarrow{x}_1, M_0, M_1, \overrightarrow{v}_0, \overrightarrow{v}_1, \overrightarrow{w}_0, \overrightarrow{w}_1)$ is as follows:

$$
\begin{aligned}
CT_{\vec{r}} &= \text{Encrypt}\left(PK, \vec{r}, g_2^{d'}\right) \\
&= (g^{s_2}, g_1^{s_1}, \{W_{1,i}^{s_1} \cdot F_{1,i}^{s_2} \cdot U_1^{r_i s_3}, W_{2,i}^{s_1} \cdot F_{2,i}^{s_2} \cdot U_2^{r_i s_3}\}_{i=1}^n, \\
&\quad \{T_{1,i}^{s_1} \cdot H_{1,i}^{s_2} \cdot V_1^{r_i s_4}, T_{2,i}^{s_1} \cdot H_{2,i}^{s_2} \cdot V_2^{r_i s_4}\}_{i=1}^n, \Lambda^{-s_2} g_2^{d'}) \in \mathbb{G}^{4n+2} \times \mathbb{G}_{\mathbb{T}}, \\
\hat{CT} &= e(g^{s_2}, g_2) \cdot e(g,g)^{\Omega(\lambda_1' s_3 + \lambda_2' s_4) < \vec{x}_b, \vec{v}_b >} \cdot e(g^{-\Omega}, g_2^{d's}),
\end{aligned}
$$

where $\vec{r} = \{r_0, \ldots, r_n\} \xleftarrow{U} \mathcal{F}$ and $d' \xleftarrow{U} \mathbb{Z}_p^*$.

Game 2. Game 2 is the same as Game 1 except that the reply to challenge query for $(\vec{x}_0, \vec{x}_1, M_0, M_1, \vec{v}_0, \vec{v}_1, \vec{w}_0, \vec{w}_1)$ is as the follows:

$$
\begin{aligned}
CT_{\vec{u}} &= \text{Encrypt}\,(PK, \vec{u}, M_b) \\
&= (g^{s_2}, g_1^{s_1}, \{W_{1,i}^{s_1} \cdot F_{1,i}^{s_2} \cdot U_1^{u_i s_3}, W_{2,i}^{s_1} \cdot F_{2,i}^{s_2} \cdot U_2^{u_i s_3}\}_{i=1}^n, \\
&\quad \{T_{1,i}^{s_1} \cdot H_{1,i}^{s_2} \cdot V_1^{u_i s_4}, T_{2,i}^{s_1} \cdot H_{2,i}^{s_2} \cdot V_2^{u_i s_4}\}_{i=1}^n, \Lambda^{-s_2} M_b) \in \mathbb{G}^{4n+2} \times \mathbb{G}_{\mathbb{T}}, \\
\hat{CT} &= e(g^{s_2}, g_2) \cdot e(g,g)^{\Omega(\lambda_1' s_3 + \lambda_2' s_4) < \vec{u}, \vec{u}' >} \cdot e(g^{-\Omega}, g_2^{ds_1})),
\end{aligned}
$$

where $\vec{u}, \vec{u}' \xleftarrow{U} \mathcal{F}$. We note that $\vec{u}$ and $\vec{u}'$ are chosen uniformly and independent from $\vec{x}_b$ and $\vec{v}_b$, respectively. $CT_{\vec{w}}$ is generated as in Game 1.

Let $\text{Adv}_{\mathcal{A}}^{(0')}(\lambda)$, $\text{Adv}_{\mathcal{A}}^{(1)}(\lambda)$ and $\text{Adv}_{\mathcal{A}}^{(2)}(\lambda)$ be the advantages of $\mathcal{A}$ in Games 0, 1 and 2, respectively. We will use three lemmas (Lemmas 2, 3, 4) that evaluate the gaps between pairs of neighboring games. From these lemmas we obtain:

$$
\begin{aligned}
\text{Adv}_{\mathcal{A}}^{(0')}(\lambda) &\le |\text{Adv}_{\mathcal{A}}^{(0')}(\lambda) - \text{Adv}_{\mathcal{A}}^{(1)}(\lambda)| + |\text{Adv}_{\mathcal{A}}^{(1)}(\lambda) - \text{Adv}_{\mathcal{A}}^{(2)}(\lambda)| \\
&\quad + \text{Adv}_{\mathcal{A}}^{(2)}(\lambda) \le \text{Adv}_{\beta_{1-1}}^{IPE\text{-}AH}(\lambda) + \text{Adv}_{\beta_{1-2}}^{IPE\text{-}AH}(\lambda).
\end{aligned}
$$

Lemma 2. For any adversary $\mathcal{A}$, there exists a probabilistic machine β_{1-1} and β_{1-2}, whose running time is essentially the same as that of $\mathcal{A}$, such that for any security parameter λ, $|\text{Adv}_{\mathcal{A}}^{(0')}(\lambda) - \text{Adv}_{\mathcal{A}}^{(1)}(\lambda)| \le \text{Adv}_{\beta_{1-1}}^{IPE\text{-}AH}(\lambda) + \text{Adv}_{\beta_{1-2}}^{IPE\text{-}AH}(\lambda)$.

Proof of Lemma 2. We construct probabilistic machines β_{1-1} and β_{1-2} against the fully-attribute hiding security using an adversary $\mathcal{A}$ in a security game (Game 0' or Game 1) as a block box. To this purpose, we consider the intermediate game Game 1' that is the same as Game 0' except that $CT_{\vec{w}}$ of the reply the challenge re-encrypted ciphertext is of the form of Game 1. Hence to prove that $|\text{Adv}_{\mathcal{A}}^{(0')}(\lambda) - \text{Adv}_{\mathcal{A}}^{(1')}(\lambda)| \leq \text{Adv}_{\beta_{1-1}}^{IPE\text{-}AH}(\lambda)$, we construct a probabilistic machine β_{1-1} against the fully attribute-hiding security using the adversary $\mathcal{A}$ in a security game (Game 0' or Game 1') as block box as follows:

1. β_{1-1} plays a role of the challenger in the security game against the adversary $\mathcal{A}$.

2. β_{1-1} generates a public and secret key and provides $\mathcal{A}$ with the public key and keeps the secret key as details are stated in Section 6.

   $$PK = (g, g_1, \{W_{1,i}, W_{2,i}, F_{1,i}F_{2,i}\}_{i=1}^n, \{T_{1,i}, T_{2,i}, H_{1,i}, H_{2,i}\}_{i=1}^n,$$
   $$\{U_i, V_i\}_{i=1}^2, \Lambda)$$
   $$MSK = (\{w_{1,i}, w_{2,i}, t_{1,i}, t_{2,i}, f_{1,i}, f_{2,i}, h_{1,i}, h_{2,i}\}_{i=1}^n, \{\delta_i, \theta_i\}_{i=1}^2, g_2).$$

3. When a private key query is issued for a vector $\vec{v}$, $\beta - 1$ computes a normal form decryption key and provides $\mathcal{A}$ with $SK_{\vec{v}} = (K_A, K_B, \{K_{1,i}, K_{2,i}\}_{i=1}^n, \{K_{3,i}, K_{4,i}\}_{i=1}^n)$.

4. When a re-encryption key query is issued for $(\vec{v}, \vec{w})$, β_{1-1}, computes a normal form re-encryption key $RK_{\vec{v},\vec{w}} = (K_A', K_B', \{K_{1,i}', K_{2,i}'\}_{i=1}^n, \{K_{3,i}', K_{4,i}'\}_{i=1}^n)$ along with $CT_{\vec{w}} = (PK, \vec{w}, g_2^d)$.

5. When a re-encryption query is issued for $(\vec{v}, \vec{w}, CT_{\vec{x}})$, the challenger β_{1-1} computes a normal form of re-encryption $CT_{\vec{x}}'$ and provides $\mathcal{A}$ with $CT_{\vec{x}}' = (A, B, CT_{\vec{w}}, \hat{C}T, D)$.

6. When a challenge query is issued for $(\vec{x}_0, \vec{x}_1, M_0, M_1, \vec{v}_0, \vec{v}_1, \vec{w}_0, \vec{w}_1)$, β_{1-1} picks a bit $b \xleftarrow{U} \{0, 1\}$ and computes $CT_{\vec{x}}, CT_{\vec{w}}, CT_{\vec{x}}'$. The β_{1-1} submits $(X_b := g_2 d, X_{(1-b)} := R, \vec{x}_b, := \vec{w}_b, \vec{x}_{1-b} := \vec{r})$ to the attribute-hiding challenger underlying *IPE* scheme (See Definition 1) where R and $\vec{r}$ are chosen independently uniform. It then receives $CT_{\vec{w}\beta}$ for $\beta \xleftarrow{U} \{0, 1\}$. Finally β_{1-1} provides $\mathcal{A}$ with a challenge ciphertext $CT_b' = (A, B, CT_{\vec{w}_b} = CT_{\vec{w}_\beta}, \hat{C}T, D)$.

7. $\mathcal{A}$ finally outputs b_1. β_{1-1} outputs $\beta = 0$ if $b = b'$, otherwise outputs $\beta = 1$. Since CT' of the challenge re-encrypted ciphertext is of the form Game $0'$ (resp. Game 1 if $\beta = 0$ (resp. $\beta = 1$), the view of $\mathcal{A}$ given by β_{1-1} is distributed as Game $1'$ (resp. Game $0'$) if $\beta = 0$ (resp. $\beta = 1$). Then, $|\mathrm{Adv}_{\mathcal{A}}^{(0)}(\lambda) - \mathrm{Adv}_{\mathcal{A}}^{(1')}(\lambda)| \le |Pr[b = b'] - \frac{1}{2}|\mathrm{Adv}_{\beta_{1-1}}^{IPE\text{-}AH}(\lambda)$.

In a similar way, we construct a probabilistic machine β_{1-2} against the fully attribute-hiding security using an adversary $\mathcal{A}$ in a security game (Game $1'$ or Game 1) as a block box. Game 1 is the same as Game $1'$ except that $CT_{\vec{w}}$ of the reply to the challenge re-encrypted ciphertext $CT_{\vec{x}}$ where $\vec{r} \xrightarrow{U} \mathcal{F}$. Hence, we have $|\mathrm{Adv}_{\mathcal{A}}^{(1')}(\lambda) - \mathrm{Adv}_{\mathcal{A}|}^{(1)}(\lambda)| \le \mathrm{Adv}_{\beta_{1-2}}^{IPE\text{-}AH}(\lambda)$. Therefore, we can prove this Lemma by using hybrid argument.

Lemma 3. For any adversary $\mathcal{A}$, $\mathrm{Adv}_{\mathcal{A}}^{(1)}(\lambda) = \mathrm{Adv}_{\mathcal{A}}^{(2)}(\lambda)$.

Proof of Lemma 3. From the adversary's view $CT_{\vec{x}}$ of Game 1 and $CT_{\vec{u}}$ of Game 2 where $\vec{u} \xrightarrow{U} \mathcal{F}$ are information theoretically indistinguishable.

Lemma 4. For any adversary $\mathcal{A}$, $\mathrm{Adv}_{\mathcal{A}}^{(2)}(\lambda) = 0$.

Proof of Lemma 4. The value b is independent from adversary's view in Game 2. Hence, $\mathcal{A}$, $\mathrm{Adv}_{\mathcal{A}}^{(2)}(\lambda) = 0$.

Proof of Theorem 1 in the case $\tau_{M,\vec{x},\vec{v}=1}$

Lemma 5. The proposed *IPPRE* scheme is predicate- and attribute-hiding for re-encrypted ciphertexts against chosen-plaintext attack in the case $\tau_{M,\vec{x},\vec{v}=1}$ under the attribute hiding underlying *IPE* scheme.

For any adversary $\mathcal{A}$, there exists probabilistic mechanisms ϵ_{2-1} and ϵ_{2-2}, whose running times are essentially the same as that of $\mathcal{A}$ such that for any security parameter λ in the case $\tau_{M,\vec{x},\vec{v}=1}$.

$$\Pr[\mathcal{A}_{wins}|T_{M,\vec{x},\vec{v}} = 0] - \frac{1}{2} \le \mathrm{Adv}_{\epsilon_{2-1}}^{IPE\text{-}AH} + \mathrm{Adv}_{\epsilon_{2-2}}^{PAH\text{-}AH}.$$

The aim of game transformation here is that $CT_{\vec{w}_b}$ is changed to ciphertext with opposite attribute $\vec{w}_{(1-b)}$. Again, we employ two games Game $0'$ and Game 1. In Game 1, the $CT_{\vec{w}_b}$ is changed to Encrypt $(PK, \vec{w}_{(1-b)}, g_2{}^d)$, respectively, by using the fully attribute-hiding security of the *IPE* scheme.

Proof of Lemma 5. To prove this lemma, we consider the following games:

Game $0'$. Same as Game 0 except that flip a coin $\tau_{M,\vec{x},\vec{v}} \xleftarrow{U} \{0,1\}$ before setup, and the game is aborted if $\tau_{M,\vec{x},\vec{v}} \neq S_{M,\vec{x},\vec{v}}$. We consider the case with $\tau_{M,\vec{x},\vec{v}} = 1$. Again here we only describe the components which are changed in the other games. The reply to challenge query for $(M, \vec{x}, \vec{v}, \vec{w}_0, \vec{w}_1)$ with $(M, \vec{x}, \vec{v}) = (M_0, \vec{x}_0, \vec{v}_0) = (M_1, \vec{x}_1, \vec{v}_1)$ is:

$$CT_{\vec{w}_b} = \text{Encrypt}\left(PK, \vec{w}_b, g_2^d\right)$$
$$= (g^{s_2}, g_1^{s_1}, \{W_{1,i}^{s_1} \cdot F_{1,i}^{s_2} \cdot U_1^{w_{ib}s_3}, W_{2,i}^{s_1} \cdot F_{2,i}^{s_2} \cdot U_2^{w_{ib}s_3}\}_{i=1}^n,$$
$$\{T_{1,i}^{s_1} \cdot H_{1,i}^{s_2} \cdot V_1^{w_{ib}s_4}, T_{2,i}^{s_1} \cdot H_{2,i}^{s_2} \cdot V_2^{w_{ib}s_4}\}_{i=1}^n, \Lambda^{-s_2} g_2^d) \in \mathbb{G}^{4n+2} \times \mathbb{G}_{\mathbb{T}},$$

where $d \xleftarrow{U} \mathbb{Z}_p^*$.

Game 1. Game 1 is the same as Game $0'$ except that the reply to the challenge query for $(M, \vec{x}, \vec{v}, \vec{w}_0, \vec{w}_1)$ with $M, \vec{x}, \vec{v} = (M_0, \vec{x}_0, \vec{v}_0) = (M_1, \vec{x}_1, \vec{v}_1)$ is:

$$CT_{\vec{w}_{1-b}} = \text{Encrypt}\left(PK, \vec{w}_{1-b}, g_2^d\right)$$
$$= (g^{s_2}, g_1^{s_1}, \{W_{1,i}^{s_1} \cdot F_{1,i}^{s_2} \cdot U_1^{w_{i1-b}s_3}, W_{2,i}^{s_1} \cdot F_{2,i}^{s_2} \cdot U_2^{w_{i1-b}s_3}\}_{i=1}^n,$$
$$\left\{T_{1,i}^{s_1} \cdot H_{1,i}^{s_2} \cdot V_1^{w_{i1-b}s_4}, T_{2,i}^{s_1} \cdot H_{2,i}^{s_2} \cdot V_2^{w_{i1-b}s_4}\right\}_{i=1}^n, \Lambda^{-s_2} g_2^d\right) \in \mathbb{G}^{4n+2} \times \mathbb{G}_T.$$

Let $\text{Adv}_{\mathcal{A}}^{(0')}(\lambda)$ and $\text{Adv}_{\mathcal{A}}^{(1)}(\lambda)$ be the advantage of $\mathcal{A}$ in Game $0'$ and Game 1, respectively. In order to evaluate the gaps between pairs of neighboring games, we consider the following Lemmas (6 and 7). We have:

$$\text{Adv}_{\mathcal{A}}^{(0')}(\lambda) \leq |\text{Adv}_{\mathcal{A}}^{(0')}(\lambda) - \text{Adv}_{\mathcal{A}}^{(1)}(\lambda) + \text{Adv}_{\mathcal{A}}^{(1)}(\lambda) \leq \text{Adv}_{\beta_{2-1}}^{IPE\text{-}AH}(\lambda) +$$
$$\text{Adv}_{\beta_{2-2}}^{IPE\text{-}AH}(\lambda) + \text{Adv}_{\mathcal{A}}^{(1)}(\lambda).$$

The proof is completed from the Lemma 7 since $\text{Adv}_{\mathcal{A}}^{(0')}(\lambda) \leq \frac{1}{2}(\text{Adv}_{\beta_{2-1}}^{IPE\text{-}AH}(\lambda) + \text{Adv}_{\beta_{2-2}}^{IPE\text{-}AH}(\lambda))$.

Lemma 6. For any adversary $\mathcal{A}$, there exists a probabilistic machines β_{2-1} and β_{2-2}, whose running time is essentially the same as that of $\mathcal{A}$, such that for any security parameter λ, $|\text{Adv}_{\mathcal{A}}^{(0')}(\lambda) - \text{Adv}_{\mathcal{A}}^{(1)}(\lambda)| \leq \text{Adv}_{\beta_{2-1}}^{IPE\text{-}AH}(\lambda) + \text{Adv}_{\beta_{2-1}}^{IPE\text{-}AH}(\lambda)$.

The proof of this lemma is similar to the Lemma 2.

Lemma 7. For any adversary $\mathcal{A}$, $|\text{Adv}_{\mathcal{A}}^{(1)}(\lambda) = -\text{Adv}_{\mathcal{A}}^{(0')}(\lambda)$. The challenge re-encrypted ciphertext for the opposite bit $1 - b$ to the challenge bit b and the others components are normal forms in Game 1. Hence, success probability $Pr[Succ_{\mathcal{A}}^{(1)}]$ in Game 1 is $1 - Pr[Succ_{\mathcal{A}}^{((0'))}]$, where $Succ_{\mathcal{A}}^{(0')}$ is success probability in Game $0'$. Therefore, we have $\text{Adv}_{\mathcal{A}}^{(1)}(\lambda) = -\text{Adv}_{\mathcal{A}}^{(0')}(\lambda)$.

7 Experimental Evaluation

In this section, we present our evaluation results of the proposed inner-product proxy re-encryption (*IPPRE*) scheme in terms of computation and communication costs as well as storage overhead. We present both theoretical and the experimental results with the assumption that a total number of attributes in the system is equal to n.

7.1 Theoretical Results

The computational load, defined in terms of number of computational steps required to perform a given task, can be described in the following terms, depending on the party who is performing the task itself:

- **Computational Load of the Trusted Authority.** The trusted authority has to execute three algorithms: Setup, KeyGen and Re-KeyGen. In the Setup algorithm, the main computation overhead consists of $(8n + 5)$ exponentiation operations on the group $\mathbb{G}_1$ and one pairing operation $e(g, g_2)$ that can be ignored since it can computed in advance (pre-computed). The main computation overhead of KeyGen algorithm belongs to the private key generation, which consumes $(9n)$ exponentiation operations on the group $\mathbb{G}_2$. The Re-KeyGen algorithm requires $(12n + 2)$ exponentiation operations on the group $\mathbb{G}_1$ encrypting g_2^d and $(13n)$ exponentiation operations on the group $\mathbb{G}_2$ for generating re-encryption key.

- **Computational Load on the Data Owner.** The computational overhead on the side of the data owner is caused by the execution of the Encrypt algorithm, which needs $(12n+2)$ exponentiation operations on the group $\mathbb{G}_1$ and one exponentiation operations on the group $\mathbb{G}_T$.

- **Computational Load on the Proxy.** The proxy is responsible for transforming the ciphertext by executing the Re-Encrypt algorithm, which requires $(4n + 2)$ pairing operations.

- **Computational Load for Users.** The computational overhead on the user side is mainly caused by the Decrypt algorithm. According to our protocol, we have two Decrypt algorithms: one decrypting a ciphertext and another decrypting a re-encrypted ciphertext. The computational overhead of the former consists of $(4n + 2)$ pairing operations. The computational overhead of the latter consists of $(4n + 3)$ pairing operations.

- **Communication Load.** The original ciphertext has four parts: $A = g^{s_1}$, $B = g^{s_1 \Omega}, \{C_{1,i}, C_{2,i}\}_{i=1}^{n}$ and $\{C_{3,i}, C_{4,i}\}_{i=1}^{n}$. Each C_i has three elements. The ciphertext contains $(12n + 2)$ $\mathbb{G}_1$ and a (1) $\mathbb{G}_T$ group elements in total. The re-encrypted ciphertext contains $(4n + 2)$ $\mathbb{G}_T$ group elements.

- **Storage Load for Users.** The main storage load of each user is for the private key $SK_{\vec{v}}$, which represents $(9n)$ $\mathbb{G}_2$ group elements in total.

7.2 Experimental Results

We implemented our scheme in C using the Pairing-Based Crypto (*PBC*) library [23]. The experiments were carried out on Ubuntu *16.04 LTS* with *2.60 GHz 8x Intel(R) Core(TM) i7-4720HQ CPU* and *16 GB RAM*.

Using Different Types of Elliptic Curves. The choice of elliptic curve parameters impacts on the credential, signature sizes, and the computational efficiency. We measured the execution time of our scheme on three different types of elliptic curves with 80 bits of security level: SuperSingular (*SS*) curve (type *A*), *MNT* curves (type *D*) and Barreto-Naehrig (*BN*) curve (type *F*), respectively as defined in *PBC* [23]. The parameters of each curve are shown in Table 1.

Elliptic curves are classified into two categories: symmetric bilinear group $(e : \mathbb{G}_1 \times \mathbb{G}_2 \to \mathbb{G}_T, \mathbb{G}_1 = \mathbb{G}_2)$ and asymmetric bilinear group $(e : \mathbb{G}_1 \times \mathbb{G}_2 \to \mathbb{G}_T, \mathbb{G}_1 \neq \mathbb{G}_2)$. To achieve fast pairing computation, elliptic curves

Table 1   Curve Parameters

Type of Elliptic Curve	SuperSingular	MNT159	MNT201	BN
Bit length of q	512	159	201	158
Bit length of r	160	158	181	158
Embedding Degree	2	6	6	12
Curve	$y^2 = x^3 + x$	$y^2 = x^3 + ax + b$	$y^2 = x^3 + ax + b$	$y^2 = x^3 + b$

from symmetric bilinear groups with small embedding degree are chosen. On the other hand, elliptic curves from asymmetric bilinear groups with high embedding degree offer a good operation for short group element size. For a symmetric bilinear group, we selected the SuperSingular curve over a prime finite field with embedding degree of 2 and the base field size of $\mathbb{G}$ equal to 512 bits. Then for asymmetric bilinear groups, we considered two *MNT* curves (namely *MNT*159 and *MNT*201) with embedding degree of 6 and one *BN* curve with embedding degree of 12 and the base 2 and 3, the execution time of each algorithm of our scheme considering an increasing number of attributes from 5 to 30 over 100 runs. The execution time of each algorithm increases linearly with the number of attributes according to its computational overhead (see Section 7.1).

The computational overhead of Encrypt algorithm is dominated by exponentiation operation on group $\mathbb{G}_1$ and therefore *MNT*159 curve and *SS* curve respectively with smaller and larger base field size of $\mathbb{G}_1{}^2$ have the best and worst encryption performance. As shown in Table 3, for 5 attributes, the Encrypt algorithm takes about 19ms under *MNT*159 curve and 41ms under *SS* curve. On the other hand, the computational overhead of the KeyGen and the Re-KeyGen algorithms is dominated by exponentiation operation on group $\mathbb{G}_2$. As we can see from Table 3, the execution time of the KeyGen and the Re-KeyGen algorithms under *BN* curve that has smaller base field size of $\mathbb{G}_2{}^3$ among other curves is more efficient.

The embedding degree of elliptic curves directly influences the size of $\mathbb{G}_T$ and increases the complexity of pairing computation. Therefore, the

Table 2 Average execution time (ms) of each algorithm of the proposed *IPPRE* scheme on eliptic curves *SS* and *MNT159*

Curve	SS			MNT159		
Attribute.num	5	10	30	5	10	30
Encrypt	41.6	78.6	234	19	33.8	91.5
Keygen	54.6	108.5	318.7	157	308.7	935
Decrypt	13.4	24.9	69.5	33.2	61.5	173.6
Re-encrypt	13.5	24.8	71.1	33	61.6	176.3
Re-keygen	131.1	240.9	715.5	273.5	509.7	1497.3
Re-Decrypt	17.2	28.9	73.9	47.2	76.6	188.9

---

[2]The base field size of $\mathbb{G}_1$ *MNT*159, *BN*, *MNT*201 and *SS* curves are 159 bits, 160 bits, 201 bits and 512 bits, respectively [23].

[3]The base field size of $\mathbb{G}_2$ *BN*, *MNT*159, *SS* and *MNT*201 curves are 320 bits, 477 bits, 512 bits and 603 bits, respectively [23].

Table 3 Average execution time (ms) of each algorithm of the proposed *IPPRE* scheme on eliptic curves *MNT201* and *BN*

Curve	MNT201			BN		
Attribute.num	5	10	30	5	10	30
Encrypt	25	45.4	123.7	34	48.9	106
Keygen	205	403.4	1219	40.2	79.4	241.2
Decrypt	42.8	80	235.6	367.4	691.4	2082.6
Re-encrypt	43.7	80.6	231.2	367.3	700.3	2025.4
Re-keygen	359.7	668.2	1960.9	87.8	153.5	447.4
Re-Decrypt	63.1	100.7	250.4	380.5	711.8	2036.8

SS curve with embedding degree of 2 has the best execution time for the algorithms Re-Decrypt, Decrypt and Re-Decrypt among other curves, as we can see in Tables 2. While, the *BN* curve with embedding degree 12 has higher execution time for the Re-Decrypt, Decrypt and Re-Decrypt algorithms. Specifically, from the Tables 2 and 3 we can see that, in the case of 5 attributes the Decrypt and Re-Decrypt algorithms take less than 18ms for the *SS* curve and less than 63ms for *MNT* curves, while for 10 attributes these algorithms take about 29ms for *SS* curve and less than 100ms for *MNT* curves.

8 Conclusions

In this paper, we extend Park's inner-product encryption method [26]. We present a new inner-product proxy re-encryption (*IPPRE*) scheme for sharing data among users with different access policies via the proxy server. The proposed scheme allows updating attribute sets without re-encryption, making policy updates extremely efficient. We fulfill the security model for our *IPPRE* and show that the scheme is adaptive attribute-secure against chosen plaintext under standard Decisional Linear (*D-Linear*) assumption. Moreover, we test the execution time of each algorithm of our protocol on different types of elliptic curves. The execution times hint that our approach is the first step towards a promising direction. As a future work, we plan to show how to apply our scheme to achieve a multi-authority version [6].

References

[1] G. Ateniese, K. Fu, M. Green, and S. Hohenberger (2006). Improved proxy re-encryption schemes with applications to secure distributed storage. *ACM Trans. Inf. Syst. Secur.*, 9, 1–30.

[2] J. Bethencourt, A. Sahai, and B. Waters (2007). Ciphertext-policy attribute-based en cryption. In *Proceedings of the 2007 IEEE Symposium on Security and Privacy, SP '07*, (IEEE Computer Society: Washington, DC, USA), 321–334.

[3] M. Blaze, G. Bleumer, and M. Strauss (1998). *Divertible Protocols and Atomic Proxy Cryptography*. Springer: Berlin, Heidelberg, 127–144.

[4] D. Boneh, X. Boyen, and H. Shacham (2004). *Short Group Signatures*. Springer: Berlin, Heidelberg, pp. 41–55.

[5] D. Boneh and M. Franklin (2001). *Identity-Based Encryption from the Weil Pairing*. Springer: Berlin, Heidelberg, 213–229.

[6] M. Chase (2007). Multi-authority attribute based encryption. In *Proceedings of the 4th Theory of Cryptography Conference, TCC 2007, Amsterdam, The Netherlands*. (ed) Salil P. Vadhan, *Theory of Cryptography*, Lecture Notes in Computer Science, Vol. 4392, (Springer), pp. 515–534.

[7] L. Cheung and Calvin C (2007). Newport. Provably secure ciphertext policy ABE. *IACR Cryptology ePrint Archive*, 2007:183.

[8] T. El Gamal (1985). "A public key cryptosystem and a signature scheme based on discrete logarithms," in *Proceedings of CRYPTO 84 on Advances in Cryptology*, New York, NY, USA, (Springer-Verlag New York, Inc.), 10–18.

[9] K. Emura, A. Miyaji, A. Nomura, K. Omote, and M. Soshi (2009). *A Ciphertext-Policy Attribute-Based Encryption Scheme with Constant Ciphertext Length*. Springer: Berlin, Heidelberg, 13–23.

[10] V. Goyal, O. Pandey, A. Sahai, and B. Waters (2006). "Attribute-based encryption for fine-grained access control of encrypted data," in *Proceedings of the 13th ACM Conference on Computer and Communications Security, CCS '06*, (ACM: New York, NY, USA), 89–98.

[11] M. Green and G. Ateniese (2007). *Identity-Based Proxy Re-encryption*. Springer: Berlin, Heidelberg, 288–306.

[12] H. Guo, F. Ma, Z. Li, and C. Xia (2015). "Key-exposure protection in public auditing with user revocation in cloud storage," in *Revised Selected Papers of the 6th International Conference on Trusted Systems, INTRUST 2014*, (LNCS, Vol. 9473), 127–136.

[13] S. Guo, Y. Zeng, J. Wei, and Q. Xu (2008). Attribute-based re-encryption scheme in the standard model. *Wuhan University Journal of Natural Sciences*, 13, 621–625.

[14] J. Katz, A. Sahai, and B. Waters (2008). *Predicate Encryption Supporting Disjunctions, Polynomial Equations, and Inner Products.* Springer: Berlin, Heidelberg, 146–162.

[15] Y. Kawai and K. Takashima (2013). Fully-anonymous functional proxy-re-encryption. *IACR Cryptology ePrint Archive*, 2013:318.

[16] A. Lewko, T. Okamoto, A. Sahai, K. Takashima, and B. Waters (2010). *Fully Secure Functional Encryption: Attribute-Based Encryption and (Hierarchical) Inner Product Encryption.* Springer: Berlin, Heidelberg, 62–91.

[17] H. Li and L. Pang (2016). Efficient and adaptively secure attribute-based proxy reencryption scheme. *IJDSN*, 12, 5235714:1–5235714:12.

[18] K. Li (2013). Matrix access structure policy used in attribute-based proxy re-encryption. *CoRR*. abs/1302.6428, eprint: arXiv:1302.6428.

[19] K. Liang, L. Fang, W. Susilo, and D. S. Wong (2013). "A ciphertext-policy attribute-based proxy re-encryption with chosen-ciphertext security," in *2013 5th International Conference on Intelligent Networking and Collaborative Systems*, 552–559.

[20] K. Liang, M. H. Au, W. Susilo, D. S. Wong, G. Yang, and Y. Yu (2014). *An Adaptively CCA-Secure Ciphertext-Policy Attribute-Based Proxy Re-Encryption for Cloud Data Sharing.* Springer International Publishing: Cham, 448–461.

[21] X. Liang, Z. Cao, H. Lin, and J. Shao (2009). "Attribute based proxy re-encryption with delegating capabilities," in *Proceedings of the 2009 ACM Symposium on Information, Computer and Communications Security, ASIACCS 2009, Sydney, Australia*, 276–286.

[22] S. Luo, J. Hu, and Z. Chen (2010). "Ciphertext policy attribute-based proxy re-encryption," in *Proceedings of the 12th International Conference on Information and Communications Security, ICICS'10*, (Springer-Verlag: Berlin, Heidelberg), 401–415.

[23] B. Lynn (2007). *Pairing Based Cryptography Library.* Available at: http://crypto.stanford.edu/pbc/

[24] M. Mambo and E. Okamoto (1997). Proxy cryptosystems: Delegation of the power to decrypt ciphertexts. *IEICE Transactions on Fundamentals of Electronics, Communications and Computer Sciences*, 80A, 54–63.

[25] T. Okamoto and K. Takashima (2009). *Hierarchical Predicate Encryption for Inner-Products.* Springer: Berlin, Heidelberg, 214–231.

[26] J. H. Park (2011). Inner-product encryption under standard assumptions. *Des. Codes Cryptography*, 58, 235–257.

[27] Z. Qin, H. Xiong, S. Wu, and J. Batamuliza (2017). A survey of proxy re-encryption for secure data sharing in cloud computing. *IEEE Transactions on Services Computing*, PP(99), 1–1. doi: 10.1109/TSC.2016.2551238

[28] A. Sahai and B. Waters (2005). *Fuzzy Identity-Based Encryption.* Springer: Berlin, Heidelberg, 457–473.

[29] H. Seo and H. Kim (2012). Attribute-based proxy re-encryption with a constant number of pairing operations. *J. Inform. and Commun. Convergence Engineering*, 10, 53–60.

[30] M. Sepehri, S. Cimato, and E. Damiani (2017). "Efficient implementation of a proxy-based protocol for data sharing on the cloud," in *Proceedings of the Fifth ACM International Workshop on Security in Cloud Computing, SCC@AsiaCCS 2017,* Abu Dhabi, United Arab Emirates, April 2, 2017, pp. 67–74.

[31] M. Sepehri, S. Cimato, E. Damiani, and C. Y. Yeuny (2015). "Data sharing on the cloud: A scalable proxy-based protocol for privacy-preserving queries," in *Proceedings of the 7th IEEE International Symposium on Ubisafe Computing in Conjunction with 14th IEEE Conference on Trust, Security and Privacy in Computing and Communications, TrustCom/BigDataSE/ISPA,* Helsinki, Finland, 1357–1362.

[32] M. Sepehri, and A. Trombetta (2017). Secure and Efficient Data Sharing with Atribute-based Proxy Re-encryption Scheme. In *Proceedings of the 12th International Conference on Availability, Reliability and Security.* ACM: Reggio Calabria, Italy, 1–63.

[33] D. Thilakanathan, S. Chen, S. Nepal, and R. A. Calvo (2014). *Secure Data Sharing in the Cloud.* Springer: Berlin, Heidelberg, 45–72.

[34] B. Waters (2009). *Dual System Encryption: Realizing Fully Secure IBE and HIBE under Simple Assumptions.* Springer: Berlin, Heidelberg, 619–636.

[35] B. Waters (2011). *Ciphertext-Policy Attribute-Based Encryption: An Expressive, Efficient, and Provably Secure Realization.* Springer: Berlin, Heidelberg, 53–70.

[36] S. Yu, C. Wang, K. Ren, and W. Lou (2010). "Attribute based data sharing with attribute revocation," in *Proceedings of the 5th ACM Symposium on Information, Computer and Communications Security, ASIACCS '10,* (ACM: New York, NY, USA), 261–270.

Biographies

Masoomeh Sepehri is a Ph.D. candidate in Computer Science at the University of Milan. Her research activity mainly focused on the design of cryptographic protocols for secure data sharing in the cloud computing and Internet of Things.

Alberto Trombetta is Associate Professor at the Department of Scienze Teoriche e Applicate at Insubria University in Varese, Italy. His main research interests include security and privacy issues in data management systems, applied cryptography and trust management.

Maryam Sepehri received the Ph.D. degree in Computer Science from the University of Milan, Italy, in 2014. She is currently pursuing postdoctoral research fellow at the University of Waterloo, Canada. Her research interest includes privacy-preserving query processing over encrypted data.